FLUID MECHANICS & FLUID MACHINES

AUTHOR

Dr. R.P. Saini

Department of Hydro and Renewable Energy
Indian Institute of Technology Roorkee
Roorkee - 247 667 Uttarakhand
Email ID: rp.saini@hre.iitr.ac.in

REVIEWER

Dr. Malay Kumar Das

Professor,
Department of Mechanical Engineering (ME)
Indian Institute of Technology Kanpur
Kanpur – 208 016 Uttar Pradesh
Email ID: mkdas@iitk.ac.in

KHANNA BOOK PUBLISHING CO. (P) Ltd.
4C/4344, Ansari Road, Darya Ganj,
New Delhi, India

All India Council for Technical Education
Nelson Mandela Marg, Vasant Kunj,
New Delhi, India

Dear Readers,

To prevent the piracy, this book is secured with HIGH SECURITY HOLOGRAM on the front title cover. In case you don't find the hologram on the front cover title, please write us to at contact@khannabooks.com or whatsapp us at +91-99109 09320 and avail special gift voucher for yourself.

Specimen of Hologram on front Cover title:

Moreover, there is a SPECIAL DISCOUNT COUPON for you with EVERY HOLOGRAM.

How to avail this SPECIAL DISCOUNT:

Step 1: Scratch the hologram

Step 2: Under the scratch area, your "coupon code" is available

Step 3: Logon to www.khannabooks.com

Step 4: Use your "coupon code" in the shopping cart and get your copy at a special discount

Step 5: Enjoy your reading!

ISBN: 978-93-5538-837-7

Book Code: UG419EN

Fluid Mechanics and Fluid Machines

by Dr. R. P. Saini

First Edition: 2026

Originally Published by:

All India Council for Technical Education, Ministry of Education, GOI

Exclusively Published & Distributed by:

Khanna Book Publishing Co. (P) Ltd.

4C/4344, Ansari Road, Daryaganj,
New Delhi (INDIA)

Visit us at: **www.khannabooks.com**

Write us at: **contact@khannabooks.com**

To view complete list of books, Please scan the QR Code:

KPH

Printed in Bharat.

प्रो. म. जगदीश कुमार
अध्यक्ष
Prof. M. Jagadesh Kumar
Chairman

सत्यमेव जयते

आज़ादी का अमृत महोत्सव

अखिल भारतीय तकनीकी शिक्षा परिषद्
(भारत सरकार का एक सांविधिक निकाय)
(शिक्षा मंत्रालय, भारत सरकार)
नेल्सन मंडेला मार्ग, वसंत कुंज, नई दिल्ली–110070
दूरभाष : 011–26131498
ई–मेल : chairman@aicte-india.org

ALL INDIA COUNCIL FOR TECHNICAL EDUCATION
(A STATUTORY BODY OF THE GOVT. OF INDIA)
(Ministry of Education, Govt. of India)
Nelson Mandela Marg, Vasant Kunj, New Delhi-110070
Phone : 011-26131498
E-mail : chairman@aicte-india.org

FOREWORD

Engineers are the backbone of the modern society. It is through them that engineering marvels have happened and improved quality of life across the world. They have driven humanity towards greater heights in a more evolved and unprecedented manner.

The All India Council for Technical Education (AICTE), led from the front and assisted students, faculty & institutions in every possible manner towards the strengthening of the technical education in the country. AICTE is always working towards promoting quality Technical Education to make India a modern developed nation with the integration of modern knowledge & traditional knowledge for the welfare of mankind.

An array of initiatives have been taken by AICTE in last decade which have been accelerate now by the National Education Policy (NEP) 2020. The implementation of NEP under the visionary leadership of Hon'ble Prime Minister of India envisages the provision for education in regional languages to all, thereby ensuring that every graduate becomes competent enough and is in a position to contribute towards the national growth and development through innovation & entrepreneurship.

One of the spheres where AICTE had been relentlessly working since 2021-22 is providing high quality books prepared and translated by eminent educators in various Indian languages to its engineering students at Under Graduate & Diploma level. For the second year students, AICTE has identified 87 books at Under Graduate and Diploma Level courses, for translation in 12 Indian languages - Hindi, Tamil, Gujarati, Odia, Bengali, Kannada, Urdu, Punjabi, Telugu, Marathi, Assamese & Malayalam. In addition to the English medium, the 1044 books in different Indian Languages are going to support to engineering students to learn in their mother tongue. Currently, there are 39 institutions in 11 states offering courses in Indian languages in 7 disciplines like Biomedical Engineering, Civil Engineering, Computer Science & Engineering, Electrical Engineering, Electronics & Communication Engineering, Information Technology Engineering & Mechanical Engineering, Architecture, and Interior Designing. This will become possible due to active involvement and support of universities/institutions in different states.

On behalf of AICTE, I express sincere gratitude to all distinguished authors, reviewers and translators from different IITs, NITs and other institutions for their admirable contribution in a very short span of time.

AICTE is confident that these out comes based books with their rich content will help technical students master the subjects with factor comprehension and greater ease.

(Prof. M. Jagadesh Kumar)

Acknowledgement

The author is grateful to the authorities of AICTE, particularly Prof. M. Jagadesh Kumar, Chairman; Prof. M. P. Poonia, Vice-Chairman; Prof. Rajive Kumar, Member-Secretary and Dr. Amit Kumar Srivastava, Director, Faculty Development Cell for their planning to publish the books on **Fluid Mechanics & Fluid Machines**. I sincerely acknowledge the valuable contributions of the reviewer of the book Dr. Malay Kumar Das, Professor, IIT Kanpur, for making it students' friendly and giving a better shape in an artistic manner.

I also wish to express gratitude to my research scholars Mr. K. Bheemalingeswara Reddy, Mr. Ankush J. Hedau, Mr. Prashant Kumar and Ms. Jyoti Pal for their continuous help and support during the period of book writing.

This book is an outcome of various suggestions of AICTE members, experts and authors who shared their opinion and thought to further develop the engineering education in our country. Acknowledgements are due to the contributors and different workers in this field whose published books, review articles, papers, photographs, footnotes, references and other valuable information enriched us at the time of writing the book.

R.P. Saini
Author

Preface

The book titled "Fluid Mechanics & Fluid Machines" is an outcome of the rich experience of our teaching of basic Fluid Mechanics and Fluid Machines courses. The initiation of writing this book is to expose basic science to the engineering students, the fundamentals of fluid based systems as well as enable them to get an insight of the subject. Keeping in mind the purpose of wide coverage as well as to provide essential supplementary information, we have included the topics recommended by AICTE, in a very systematic and orderly manner throughout the book. Efforts have been made to explain the fundamental concepts of the subject in the simplest possible way.

During the process of preparation of the manuscript, we have considered the various standard text books and accordingly we have developed sections like critical questions, solved and supplementary problems etc. While preparing the different sections emphasis has also been laid on definitions and laws and also on comprehensive synopsis of formulae for a quick revision of the basic principles. The book covers all types of problems and these have been presented in a very logical and systematic manner. The gradations of those problems have been tested over many years of teaching to a wide variety of students.

Apart from illustrations and examples as required, we have enriched the book with numerous solved problems in every unit for proper understanding of the related topics. Under the common title "Fluid Mechanics & Fluid Machines" there are two parts covering different aspects and applications of basic Fluid Mechanics and Fluid Machines engineering. Out of those, the first one covers Fundamentals of Fluid Mechanics, the second one is based on Fluid Dynamics, the third one is related to Dimensional Analysis. The second part covers basic working principle, components, performance characteristics of Fluid Machines and their selection under different operating conditions

As far as the present book is concerned, "Fluid Mechanics & Fluid Machines" is meant to provide a thorough grounding in applied fluid mechanics and fluid machines on the topics covered. This book will prepare engineering students to apply the knowledge of Fluid Mechanics in the field and application of Fluid Machines to tackle the engineering challenges and address the related aroused questions. The subject matters are presented in a constructive manner so that an Engineering degree prepares students to work in different sectors or in national laboratories at the very forefront of technology.

We sincerely hope that the book will inspire the students to learn and discuss the ideas behind basic principles of Fluid Mechanics and Fluid Machines and will surely contribute to the development of a solid foundation of the subject. We would be thankful to all beneficial comments and suggestions which will contribute to the improvement of the future editions of the book. It gives us immense pleasure to place this book in the hands of the teachers and students. It was indeed a big pleasure to work on different aspects covering in the book.

R.P. Saini
Author

Outcome Based Education

For the implementation of an outcome based education the first requirement is to develop an outcome based curriculum and incorporate an outcome based assessment in the education system. By going through outcome based assessments evaluators will be able to evaluate whether the students have achieved the outlined standard, specific and measurable outcomes. With the proper incorporation of outcome based education there will be a definite commitment to achieve a minimum standard for all learners without giving up at any level. At the end of the programme running with the aid of outcome based education, a student will be able to arrive at the following outcomes:

PO1. Engineering knowledge: Apply the knowledge of mathematics, science, engineering fundamentals, and an engineering specialization to the solution of complex engineering problems.

PO2. Problem analysis: Identify, formulate, review research literature, and analyze complex engineering problems reaching substantiated conclusions using first principles of mathematics, natural sciences, and engineering sciences.

PO3. Design / development of solutions: Design solutions for complex engineering problems and design system components or processes that meet the specified needs with appropriate consideration for the public health and safety, and the cultural, societal, and environmental considerations.

PO4. Conduct investigations of complex problems: Use research-based knowledge and research methods including design of experiments, analysis and interpretation of data, and synthesis of the information to provide valid conclusions.

PO5. Modern tool usage: Create, select, and apply appropriate techniques, resources, and modern engineering and IT tools including prediction and modeling to complex engineering activities with an understanding of the limitations.

PO6. The engineer and society: Apply reasoning informed by the contextual knowledge to assess societal, health, safety, legal and cultural issues and the consequent responsibilities relevant to the professional engineering practice.

PO7. Environment and sustainability: Understand the impact of the professional engineering solutions in societal and environmental contexts, and demonstrate the knowledge of, and need for sustainable development.

PO8. Ethics: Apply ethical principles and commit to professional ethics and responsibilities and norms of the engineering practice.

PO9. Individual and team work: Function effectively as an individual, and as a member or leader in diverse teams, and in multidisciplinary settings.

PO10. Communication: Communicate effectively on complex engineering activities with the engineering community and with society at large, such as, being able to comprehend and write effective reports and design documentation, make effective presentations, and give and receive clear instructions.

PO11. Project management and finance: Demonstrate knowledge and understanding of the engineering and management principles and apply these to one's own work, as a member and leader in a team, to manage projects and in multidisciplinary environments.

PO12. Life-long learning: Recognize the need for, and have the preparation and ability to engage in independent and life-long learning in the broadest context of technological change.

Course Outcomes

After completion of the course the students will be able to:

CO-1: Describe fluid and its properties, continuity equation, momentum equation and Bernoulli's equation with its applications

CO-2: Explain fluid flow through channels, concept of Couette flow, Poisuielle flow and boundary layer, laminar flow through circular conduits and annuli, derivation of Darcy Weisbach equation and friction factor and Mood's diagram

CO-3: Need for dimensional analysis and its methods, similitude and its types, dimensionless parameters, model analysis and problems including application of dimensionless parameters

CO-4: Apply Euler's equation to the fluid motion, theory of rotodynamic machines, working principle of pumps and derive the work done by drawing velocity triangles, performance curves, cavitation, selection of pumps and concept of pump as turbine

CO-5: Define working principle of water turbine and its classification, derive the performance with the help of velocity triangle, performance characteristics, governing of water turbines, similarity law and selection of water turbines by applying concept of specific speed

CO-6: Analyze different types of fluid mechanics and fluid machines problems in real life

Course Outcomes	**Expected Mapping with Programme Outcomes** *(1- Weak Correlation; 2- Medium correlation; 3- Strong Correlation)*											
	PO-1	PO-2	PO-3	PO-4	PO-5	PO-6	PO-7	PO-8	PO-9	PO-10	PO-11	PO-12
CO-1	3	2	2	2	1	-	-	-	-	-	-	-
CO-2	3	1	2	1	-	-	-	-	-	-	-	-
CO-3	3	3	2	1	-	-	-	-	-	-	-	-
CO-4	3	3	3	2	1	-	-	-	-	-	-	-
CO-5	3	3	2	1	1	-	-	-	-	-	-	-
CO-6	3	3	3	2	1	-	-	-	-	-	-	-

Guidelines for Teachers

To implement Outcome Based Education (OBE) knowledge level and skill set of the students should be enhanced. Teachers should take a major responsibility for the proper implementation of OBE. Some of the responsibilities (not limited to) for the teachers in OBE system may be as follows:

- Within reasonable constraint, they should manoeuvre time to the best advantage of all students.
- They should assess the students only upon certain defined criterion without considering any other potential ineligibility to discriminate them.
- They should try to grow the learning abilities of the students to a certain level before they leave the institute.
- They should try to ensure that all the students are equipped with the quality knowledge as well as competence after they finish their education.
- They should always encourage the students to develop their ultimate performance capabilities.
- They should facilitate and encourage group work and team work to consolidate newer approach.
- They should follow Blooms taxonomy in every part of the assessment.

Bloom's Taxonomy

Level	Teacher should Check	Student should be able to	Possible Mode of Assessment
Create	Students ability to create	Design or Create	Mini project
Evaluate	Students ability to justify	Argue or Defend	Assignment
Analyse	Students ability to distinguish	Differentiate or Distinguish	Project/Lab Methodology
Apply	Students ability to use information	Operate or Demonstrate	Technical Presentation/ Demonstration
Understand	Students ability to explain the ideas	Explain or Classify	Presentation/Seminar
Remember	Students ability to recall (or remember)	Define or Recall	Quiz

Guidelines for Students

Students should take equal responsibility for implementing the OBE. Some of the responsibilities (not limited to) for the students in OBE system are as follows:

- Students should be well aware of each UO before the start of a unit in each and every course.
- Students should be well aware of each CO before the start of the course.
- Students should be well aware of each PO before the start of the programme.
- Students should think critically and reasonably with proper reflection and action.
- Learning of the students should be connected and integrated with practical and real life consequences.
- Students should be well aware of their competency at every level of OBE.

Abbreviations and Symbols

List of Abbreviations

General Terms			
Abbreviations	**Full form**	**Abbreviations**	**Full form**
SI	System International	CGS	Centimeter gram second
MKS	Meter kilogram second	NPSH	Net positive suction head
FPS	Foot pound second	PAT	Pump as turbine
Units Used			
Abbreviations	**Full form**	**Abbreviations**	**Full form**
J	Joule	N	Newton
W	Watt	r.p.m	rotational per minute

List of Symbols

Symbols	Description	Symbols	Description
ρ	Mass density	N	Speed
m	Mass	S	Specific gravity
γ	Specific weight	W	Weight
g	Acceleration due to gravity	τ	Shear stress
v	Velocity	μ	Dynamic viscosity
F	Force	A	Area
ν	Kinematic viscosity	P	Power
p	Pressure	Re	Reynolds number
D	Diameter	Fr	Froude number
V	Volume	K	Bulk modulus
dp	Change in pressure	k	Ratio of specific heats
dV	Change in volume	σ	Surface tension
R	Radius	L	Length
Q	Discharge	p	Pressure intensity
H	Head	N	Speed
C_c	Coefficient of contraction	n	Manning coefficient
C_d	Coefficient of discharge	h_L	Head losses
C_v	Co-efficient of velocity	h_f	Frictional head losses
f	Friction factor	δ_E	Energy thickness

δ^*	Displacement thickness	θ	Momentum thickness
T	Time	a	acceleration
Re	Reynolds number	N_s	Specific speed
F_i	Inertia force	F_p	Pressure force
F_v	Viscous force	F_e	Elastic force
F_g	Gravitational force	F_s	Surface tension
We	Weber number	Fr	Froude number
η	Efficiency	Eu	Euler number
τ	Shear stress	T	Torque
H_m	Manometric head	η_h	Hydraulic efficiency
h_s	Suction head	η_m	Mechanical efficiency
H_s	Static head	K_u	Speed ratio
η_d	Draft tube efficiency	ψ	Flow ratio

List of Figures

List of Tables

CONTENTS

1 Fundamentals Fluid Mechanics

UNIT SPECIFICS

Through this unit we have discussed the following aspects:

- *Definition of fluid and properties of fluids;*
- *Newton's law of viscosity and units and dimensions;*
- *Continuity equation;*
- *Momentum equation;*
- *Bernoulli's equation;*
- *Applications of Bernoulli's equation.*

The practical applications of the topics are discussed for generating further curiosity and creativity as well as improving problem solving capacity.

Besides giving a large number of multiple choice questions as well as questions of short and long answer types marked in two categories following lower and higher order of Bloom's taxonomy, assignments through a number of numerical problems, a list of references and suggested readings are given in the unit so that one can go through them for practice. It is important to note that for getting more information on various topics of interest some QR codes have been provided in different sections which can be scanned for relevant supportive knowledge.

RATIONALE

This fundamental unit on fluid mechanics helps students to get a primary idea about the properties of fluids which will help to discuss and derive various principles which will be applicable to practical devices. It explains Newton's law of viscosity, continuity equation, momentum equation and Bernoulli's equation. All these are discussed at length to develop the subject. Some related problems are pointed out with an extension to understand the working of the systems in real world which can help further for getting a clear idea of the concern topics on fluid mechanics. Fluid Mechanics is an important branch of science that essentially deals with kinematics, dynamics and statics of fluids. This permits one to analyze the operation of many day-to-day familiar phenomena around us.

PRE-REQUISITES

Mathematics: Calculus, Trigonometric, Vectors (Class XII)
Physics: Mechanics (Class XII)

UNIT OUTCOMES

List of outcomes of this unit is as follows:

U1-O1: Describe the fluid and properties of fluids
U1-O2: Describe the Newton's law of viscosity
U1-O3: Explain Continuity equation and momentum equation
U1-O4: Explain Bernoulli's equation
U1-O5: Apply Bernoulli's equation to solve its applications

Unit-1 Outcomes	EXPECTED MAPPING WITH COURSE OUTCOMES (1- Weak Correlation; 2- Medium correlation; 3- Strong Correlation)					
	CO-1	CO-2	CO-3	CO-4	CO-5	CO-6
U1-O1	3	3	3	-	3	1
U1-O2	1	1	2	2	1	-
U1-O3	2	1	3	1	2	1
U1-O4	-	-	3	1	2	2
U1-O5	3	3	3	-	3	1

1.1 DEFINITION OF FLUID

Fluid Mechanics is a science which deals with kinematics, dynamics and statics of fluid motion. A fluid may be defined as a substance which is capable of flowing and has no definite shape. Fluids are the substances where molecules are spaced and free to move and are classified into liquids and gases. In case of liquid, intermolecular distance is smaller than gases where molecules are more freedom to move and occupy the entire volume of the container. While, liquid is the fluid, which occupy a definite volume of the container and properties varies slightly with temperature and pressure. Viscosity and surface tension are the main properties of fluids based on fluids are classified. Gases have low viscosity in comparison of liquids for a given temperature. Liquids are further classified as ideal and real fluids. Fluids which have no viscosity are termed as ideal fluids, however, in nature no fluid has zero viscosity. Therefore, fluids having low viscosity may be considered as ideal fluids. Compressibility is another property depends on density of a fluid, which also defines fluids as compressible and incompressible.

1.2 PROPERTIES OF FLUIDS

The property of fluids can be intensive or extensive. Temperature, pressure and density of fluids are considered as intensive properties while mass, momentum and volume are extensive properties of the fluid. The important properties of fluids are discussed in the following sub-sections.

1.2.1 Mass Density

The mass density (ρ) of a fluid can be defined as the ratio of the mass of a fluid to its volume. It is expressed mathematically as

$$\text{Mass density}, \rho = \frac{\text{Mass of fluid}, m}{\text{Volume of fluid}, V}$$

In S.I., unit of mass density is kg/m^3.

If the density of fluid varies with pressure, then fluid is known as compressible fluid. While, for incompressible fluids density remains constant under pressure. Gases are compressible fluids and liquids are mostly incompressible fluids.

1.2.2 Specific Volume

Specific volume of a fluid is the volume of fluid per unit mass which is also reciprocal of mass density and it is expressed as

$$\text{Specific volume} = \frac{\text{Volume of fluid}}{\text{Mass of fluid}}$$

$$= \frac{1}{\frac{\text{Mass of fluid}}{\text{Volume of fluid}}} = \frac{1}{\rho}$$

It is mainly applied to gases.

In S.I., unit of specific volume is m^3/kg.

1.2.3 Specific Weight

Specific weight (γ) is defined as the ratio of weight of fluid and its volume occupied.

It can be expressed as

$$\gamma = \frac{\text{Weight of liquid}}{\text{Volume of liquid}}$$

$$\gamma = \frac{mg}{V} = \rho g \tag{1.1}$$

In S.I., unit of specific weight is N/m^3.

1.2.4 Specific Gravity

Specific gravity (S) is a dimensionless quantity; it is expressed in terms of weight density of water in case of liquids and in terms of weight density of air for gases.

It can be expressed as

$$S(\text{ for liquids}) = \frac{\text{Weight density (density) of liquid}}{\text{Weight density (density) of water}}$$

$$S(\text{ for gases }) = \frac{\text{Weight density (density) of gas}}{\text{Weight density (density) of air}}$$

SCAN ME

For better understanding properties of fluids

EXAMPLE 1.1

Example 1.1 Determine the weight, density, and specific weight of one litre of gasoline having specific gravity of 0.7.

Solution

Given data:

Specific gravity, $S = 0.7$

Volume of gasoline, $V = 1$ lit $= 0.001\ m^3$

Density, ρ is given by

$$\rho = S \times 1000$$
$$= 0.7 \times 1000$$
$$= 700\ kg/m^3. \textbf{ Ans.}$$

Specific weight, γ is given by

$$\gamma = \rho \times g$$
$$= 700 \times 9.81$$
$$= 6867\ N/m^3. \textbf{ Ans.}$$

Weight, W is given by

$$W = \gamma \times V$$
$$W = 6867 \times 0.001 = 6.867\ N. \textbf{ Ans.}$$

Example 1.2 Determine the following values of a liquid for 1 litre if weight of liquid is 7 N.

(i) specific weight, (ii) density, and (iii) specific gravity

Solution

Given data:

Volume, $V = 1$ lit $= 0.001\ m^3$

Weight of liquid, $W = 7$ N

Specific weight, γ is given by

$$\gamma = \frac{W}{V}$$

$$= \frac{7}{0.001}$$

$$= 7000\ \mathrm{N/m^3}. \textbf{ Ans.}$$

Here Density, ρ is given by

$$\rho = \frac{\gamma}{g}$$

$$= \frac{7000}{9.81}$$

$$= 713.5\ \mathrm{kg/m^3}. \textbf{ Ans.}$$

Specific gravity for liquid is given by

$$S = \frac{\text{Density of liquid}}{\text{Density of water}}$$

$$= \frac{713.5}{1000}$$

$$= 0.7135. \textbf{ Ans.}$$

EXAMPLE 1.2

1.2.5 Viscosity

As discussed above fluids are classified based on the viscosity which is considered as one of the most important property of fluids. It can be defined on the concept of moment of different layers of fluids. When a layer of the fluid moves over the adjacent layer, a resistance occurs between two layers due to which the moment of layers gets retarded. This resistive force between the layers can be explained by Newton's law of viscosity as discussed in the following section.

1.3 NEWTON'S LAW OF VISCOSITY

In order to discuss Newton's law of viscosity, let us consider two layers of a fluid apart a distance of 'dy' moving one over the other at different velocities as v and $v + dv$ as shown in Fig. 1.1.

There is a shear stress, τ acting between two adjacent layers. As per the Newton's law of viscosity, shear stress is proportional to the rate of change of velocity with respect to y. It can be expressed as;

$$\tau \propto \frac{dv}{dy}$$

or $$\tau = \mu \frac{dv}{dy} \qquad (1.2)$$

where,

μ is proportionality constant which is known as the dynamic viscosity or viscosity.

$\frac{dv}{dy}$ is the rate of shear strain.

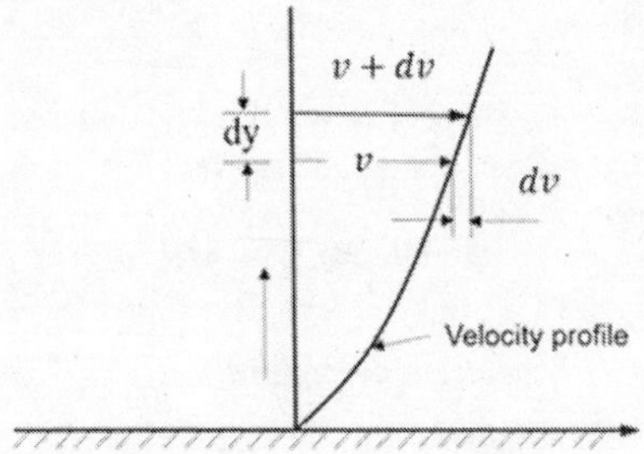

Fig. 1.1: Velocity variation near boundary

The above Eq. 1.2 can be written as

$$\mu = \frac{\tau}{\left(\frac{dv}{dy}\right)} \qquad (1.3)$$

It is therefore, viscosity can also be defined as the shear stress required to produce unit rate of shear strain.

Fluids which follow Newton's law of viscosity are known as Newtonian fluids and the others are called non-Newtonian fluids.

Unit of viscosity in S.I. is obtained as

$$\mu = \frac{\tau}{\left(\frac{dv}{dy}\right)}$$

$$= \frac{N/m^2}{\left(\frac{m}{sec}\right)/m} = \text{Ns/m}^2$$

The unit of dynamic viscosity is also expressed as poise.

1 poise = 10^{-1} Ns/m^2

Viscosity can also be written in terms of kinematic viscosity which is defined as the ratio of dynamic viscosity to the density of fluid and expressed as;

$$\nu = \frac{\text{Viscosity}}{\text{Density}}$$

$$\nu = \frac{\mu}{\rho} \qquad (1.4)$$

The units of kinematic viscosity in S.I. is obtained as

$$\nu = \frac{\text{Ns}/m^2}{kg/m^3} \quad = \frac{m^2}{\text{sec}}$$

The unit of kinematic viscosity is also expressed as stoke.

1 stoke = 10^{-4} m^2/sec

Example 1.3 Determine the force required to pull a flat plate having a surface area of 1 m^2 over another plate moving with a relative speed of 0.3 m/s if the distance between two plates is 0.1 mm. Consider the viscosity of the fluid between two plates is 1 poise. Also, calculate the corresponding power.

Solution

Given data:

Area, A = 1 m^2

Relative velocity, dv = 0.3 m/sec

Distance between two plates, dy = 0.1 mm = 0.1×10^{-3} m

Viscosity of fluid, μ = 1 poise = 0.1 Ns/m^2

Applying Newton's law of viscosity

Shear stress, $\tau = \mu \frac{dv}{dy}$

$= \frac{0.1 \times 0.3}{0.1 \times 10^{-3}}$

= 300 N/m^2

Shear force, F is given by

$F = \tau \times A$

= 300 × 1 = 300 N. **Ans.**

Power, P required to move the plate is given by

$P = F \times dv$

= 300 × 0.3 = 90 W **Ans.**

EXAMPLE 1.3

Example 1.4 Two parallel plates are separated by a fluid with a distance of between them as 0.03 mm. One plate is moving for the another fixed plate with a speed of 0.5 m/sec. Determine the viscosity of fluid if a force per unit area of 1.8 N/m^2 is required to maintain the speed of the plate.

Solution.

Given data:

Distance between two plates, dy = 0.03 mm

= 0.03×10^{-3} m

Relative speed, du = 0.5 m/s

Shear force per unit area, τ = 1.8 N/m^2

Let the fluid viscosity between the plates is μ.

EXAMPLE 1.4

EXAMPLE 1.4

The value of shear stress, τ is given by

$$\tau = \mu \frac{dv}{dy}$$

$$1.8 = \mu \frac{0.5}{0.03 \times 10^{-3}}$$

$\therefore$ $$\mu = 1.03 \times 10^{-5} \frac{\text{Ns}}{\text{m}^2}. \textbf{ Ans.}$$

EXAMPLE 1.5

Example 1.5 Determine the shear stress for a shaft having a diameter of 120 mm and rotating with 200 r.p.m in a journal bearing. If an oil having viscosity of 1 poise is used to lubricate the bearing with a clearance as 1.25 mm between shaft and journal bearing.

Solution

Given data:

Diameter of shaft, D = 120 mm = 0.12 m

Speed of shaft, N = 200 r.p.m

Viscosity of oil, μ = 1 poise = 0.1 Ns/m^2

Clearance, dy = 1.25 mm = 1.25 x10^{-3} m

Tangential speed, u of shaft is given by

$$u = \frac{\pi D N}{60}$$

$$= \frac{\pi \times 0.12 \times 200}{60} = 1.25 \text{ m/s}$$

$$du = 1.25 - 0$$

$$= 1.25 \text{ m/sec}$$

Shear stress, τ is given by

$$\tau = \mu \frac{du}{dy}$$

$$\tau = 0.1 \times \frac{1.25}{1.25 \times 10^{-3}}$$

or, $$= 100 \text{ N/m}^2. \textbf{ Ans.}$$

Example 1.6 If a fluid having specific gravity as 2.2 with kinematic viscosity of 5 stokes. Find out the viscosity of the liquid.

Solution

Given data:

Specific gravity, $S = 2.2$

Kinematic viscosity, $\nu = 5$ stokes

or, $= 5\ \text{cm}^2/\text{s}$

or, $= 5 \times 10^{-4}\ \text{m}^2/\text{s}$

Now, specific gravity, S is given by

$$S = \frac{\text{Density of liquid}}{\text{Density of water}}$$

$$2.2 = \frac{\text{Density of liquid}}{1000}$$

or, density of liquid (ρ) = $1000 \times 2.2 = 2200\ \text{kg/m}^3$

Kinematic viscosity is given by

$$\nu = \frac{\mu}{\rho}$$

$\therefore$ $\mu = 5 \times 10^{-4} \times 2200$ s

or, $= 1.1\ \text{Ns/m}^2$

or, $= 11$ poise. **Ans.**

EXAMPLE 1.6

Example 1.7 Find out the shear stress created by a fluid having dynamic viscosity of 10 poise and flowing over a flat plate at a distance of 0.2 m. If the velocity of the fluid is represented by v (m/sec) at a distance of y (m) from the plate. The velocity distribution is given as $v = 0.5y - y^2$. at a distance in which u is the velocity in metre per second at a distance y metre above the plate.

Solution

Given data:

Viscosity of fluid, $\mu = 10$ poise = 1 Ns/m2

Velocity distribution, $v = 0.5y - y^2$

Distance, $y = 0.2$ m

now,

$$\frac{dv}{dy} = 0.5 - 2y$$

at $y = 0.2$

$$\frac{dv}{dy} = 0.5 - 2 \times 0.2 = 0.5 - 0.40 = 0.1$$

EXAMPLE 1.7

EXAMPLE 1.7

Shear stress, τ is given by

$$\tau = \mu \frac{dv}{dy}$$

$$= 1 \times 0.1 \frac{\text{N}}{\text{m}^2}$$

or, $= 0.1 \frac{\text{N}}{\text{m}^2}$. **Ans.**

1.4 COMPRESSIBILITY

Compressibility of fluid is the measure of elasticity which is represented by bulk modulus of elasticity, K which is the ratio of compressive stress to volumetric strain. The concept of compressibility can be discussed by considering a cylinder fitted with a piston as shown in Fig. 1.2.

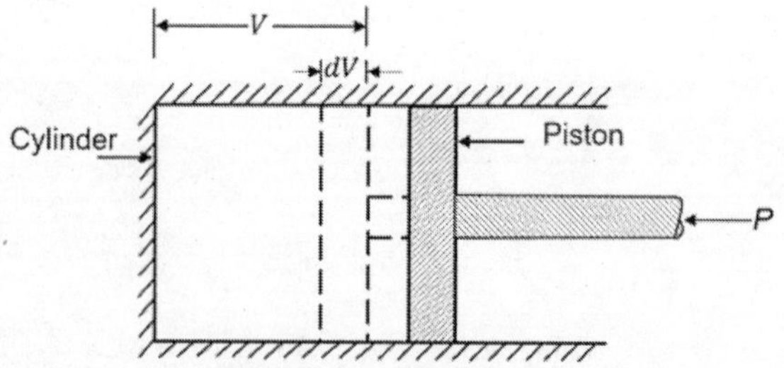

Fig 1.2: Cylinder fitted with a piston

As the pressure increases from p to $p + dp$ then volume will change from V to $V - dV$.

Thus, decrease in volume as dV corresponding to increase in pressure of dp.

Then,

$$\text{volumetric strain} = -\frac{\text{dV}}{\text{V}}$$

As per the definition of Bulk modulus, K

$$\text{Bulk modulus, K} = \frac{\text{Increase of pressure(compressive stress)}}{\text{Volumetric strain}}$$

$$\text{K} = = \frac{dp}{\frac{-\text{dV}}{\text{V}}} = \frac{-dp}{\text{dV}}\text{V} \tag{1.5}$$

Compressibility of fluid is defined as the reciprocal of the bulk modulus and expressed as

$$\text{Compressibility} = \frac{1}{K}$$

For two different processes, the relationship between bulk modulus and pressure is discussed as

(i) For isothermal process

$$p\text{V} = \text{Constant} \left\{\because \text{V} = \frac{1}{\rho}\right\} \tag{1.6}$$

By differentiating the equation (1.6)

$$pdV + Vdp = 0 \tag{1.7}$$

$$p = \frac{-Vdp}{dV} \tag{1.8}$$

Substituting this value in equation (1.5),

$$K = p \tag{1.9}$$

(ii) For adiabatic process

$$pV^k = \text{Constant} \tag{1.10}$$

Where, k is the ratio of specific heats.

By differentiating the equation (1.10),

$$pd(V^k) + V^k(dp) = 0$$

$$pkdV + Vdp = 0$$

$$pk = -\frac{Vdp}{dV}$$

Substituting this value in Eq. (1.5),

$$K = pk \tag{1.11}$$

Example 1.8 The pressure of a liquid increases from 60 N/cm^2 to 100 N/cm^2 and its volume decreases by 10%. Calculate bulk modulus of elasticity of this liquid.

Solution

Given data:

Initial pressure, $p_1 = 60\ N/cm^2$

Final pressure, $p_2 = 100\ N/cm^2$

Decrease in volume $= -\frac{dV}{V}$

$= 0.10$

Therefore, increase in pressure is given by,

$$dp = 100 - 60 = 60\ N/cm^2$$

Bulk modulus, K is given by

$$K = \frac{dp}{-\frac{dV}{V}} = \frac{60}{0.1}$$

or, $= 600\ N/cm^2$

or, $= 6$ MPa. **Ans.**

EXAMPLE 1.8

1.5 SURFACE TENSION

There are two types of forces which act between molecules of a matter; one is directly proportional to the product of molecule masses and another is inversely proportional to the distance between their centre of masses. Surface tension is the property of the liquid surface film to exert a tensile force. Due to attraction of molecules below the liquid surface, a thin layer film is formed which is in tension and a small load is supported by the lower layers of the liquid this force of molecules acts normal to the liquid surface which results in curvature of the liquid surface. In order to discuss the phenomena of surface tension an example of liquid droplet and liquid jet can be considered as illustrated in Fig. 1.3.

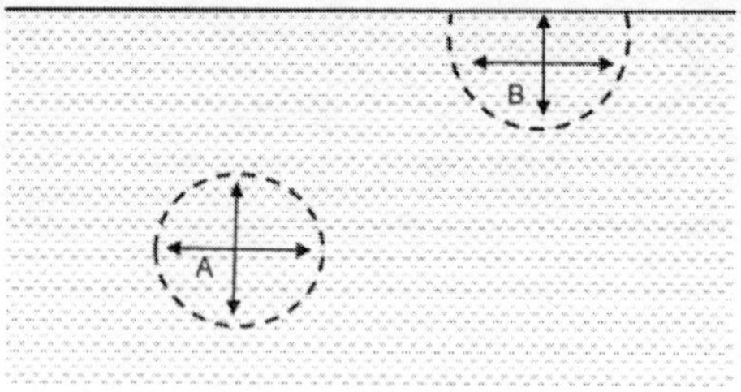

Fig. 1.3: Concept of surface tension phenomena

Let us consider a liquid molecule inside the liquid and at the surface of the liquid as represented by 'A' and 'B' respectively. The surface of the molecule will be attracted by surrounding molecule inside the liquid. The forces acting over the entire surface are equal at molecule 'A'. However, at the surface molecule 'B' there will be a difference in the forces acting above and below the liquid surfaces. The force acting below the liquid surface is more hence a curvature is seen due to surface tension. To estimate the pressure intensity due to surface tension three different cases as (i) pressure intensity inside a droplet (ii) pressure intensity inside a soap bubble and (iii) pressure intensity inside liquid jet are considered.

Case 1: let us consider a liquid droplet in spherical shape with a radius, R at the liquid surface having an internal pressure intensity of p. At the liquid surface, the droplet will look like half of the sphere therefore the projected area will be as (πR^2) over which pressure intensity p will act. However, the tensile force will be acting over surface area of the droplet (πR). These two forces will be in equilibrium and hence can be written as;

$$p\,\pi R^2 = 2\sigma\,(\pi R) \quad (1.12)$$

Therefore, pressure intensity inside a droplet can be obtained as

$$p = \frac{2\sigma}{R} \quad (1.13)$$

Case 2: In case of a soap bubble there will be two surfaces in contact with air. Both inside and outside surfaces will contribute the same amount of tensile force due to surface tension. The tensile force due to surface tension over a hemi-spherical section of soap bubble of radius R will be expressed as $2\sigma\,(2\pi R)$. An equal force due to pressure force will be acting on the hemi-spherical section as considered in the case of droplet. These two forces will be equilibrium and can be expressed as

$$p\,\pi R^2 = 2\sigma\,(2\pi R) \quad (1.14)$$

Therefore, pressure intensity inside a soap bubble can be obtained as

$$p = \frac{4\sigma}{R} \tag{1.15}$$

Case 3: Let us consider a liquid jet having radius as R, length as L and experiences a pressure intensity as p. The forces will be acting on one-half of the jet if it is cut into two halves. In this case the projected area will be $2RL$ and tensile force due to surface tension will be acting along the two sides of the jet i.e., $2L$. These two forces will be in equilibrium and expressed as follows

$$p\,(2RL) = \sigma\,(2L) \tag{1.16}$$

Therefore, pressure intensity inside liquid jet can be obtained as

Or $$p = \frac{\sigma}{R} \tag{1.17}$$

Example 1.9 Find out diameter of a water droplet assuming surface tension between air and water as 0.075 N/m at 20°C, if the difference between internal and external pressure is 0.015 N/cm^2.

Solution

Given data:

Surface tension, σ = 0.075 N/m

Pressure, p = 0.015 N/cm^2 = 0.015 x 10^4 N/m^2

Pressure intensity in the droplet is given by

$$p = \frac{4\sigma}{d}$$

Therefore, diameter, d of the droplet is given by

$$d = \frac{4\sigma}{p}$$

$$= \frac{4 \times 0.075}{0.015 \times 10^4}$$

or, d = 0.002 m = 2 mm. **Ans.**

EXAMPLE 1.9

1.6 CONTINUITY EQUATION

Based on law of conservation of mass, it can be stated that the total mass of fluid flowing through a passage will be same at different cross sections of the passage if there are no energy losses through the passage. Considering two locations of a pipe having different cross-sections as shown in Fig. 1.4 the continuity equation can be derived.

Let

m_1 and m_2 are the mass of fluid at location 1 and 2 respectively.

v_1 and v_2 are the velocity of fluid at location 1 and 2 respectively.

ρ_1 and ρ_2 are the density of fluid at location 1 and 2 respectively.

A_1 and A_2 are the cross-sectional area of passage at location 1 and 2 respectively.

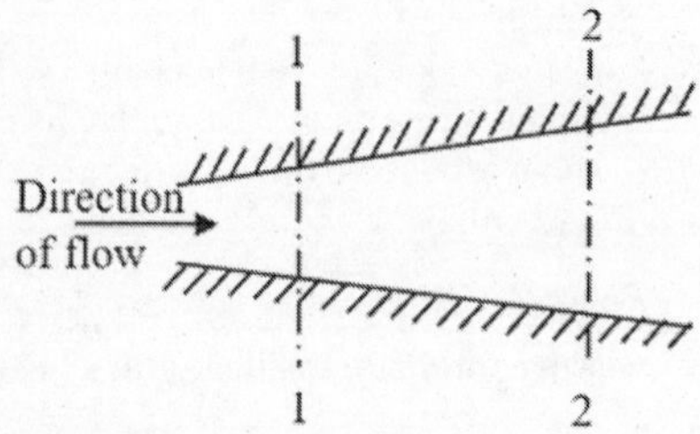

Fig. 1.4: Different locations at fluid flowing through a passage

Therefore,

The flow rate at location 1-1 = $\rho_1 A_1 v_1$

Flow rate at location 2-2 = $\rho_2 A_2 v_2$

Assuming Ideal fluid and steady flow, law of conservation of mass can be expressed as

Flow rate at location 1-1 = Flow rate at location 2-2 = constant

$$\rho_1 A_1 v_1 = \rho_2 A_2 v_2 \qquad (1.18)$$

For incompressible fluid ($\rho_1 = \rho_2$), the above equation can be expressed as

$$A_1 v_1 = A_2 v_2 \qquad (1.19)$$

Eq. 1.8 and Eq. 1.19 represent the continuity equations for compressible fluid and incompressible fluid respectively under the assumption of fluid is ideal and flow is in steady state.

EXAMPLE 1.10

Example 1.10 A fluid is flowing through an expansion pipe having inlet and outlet diameters as 150 mm and 200 mm respectively as shown in figure. If the velocity of fluid is 3.5 m/s at inlet, then find out the velocity at outlet of the pipe.

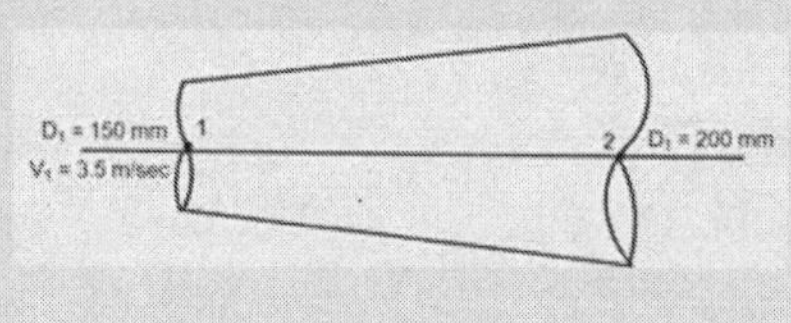

Solution:

Given data:

Velocity at point 1, v_1= 3.5 m/s

Inlet diameter, $D_1 = 150 \text{ mm} = 0.15 \text{ m}$

Outer diameter $D_2 = 200 \text{ mm} = 0.2 \text{ m}$

Area at inlet is given by

$$A_1 = \frac{\pi}{4}(D_1^2)$$

or, $$= \frac{\pi}{4}(0.15)^2 = 0.01767 \text{ m}^2$$

Area at outlet is given by

$$A_2 = \frac{\pi}{4}(D_2^2)$$

or, $$= \frac{\pi}{4}(0.2)^2 = 0.0314 \text{ m}^2$$

Applying continuity equation

$$Q = A_1 v_1 = A_2 v_2$$

∴ Velocity at outlet, V_2 is given by

$$v_2 = \frac{A_1 \times v_1}{A_2}$$

$$= \frac{0.01767 \times 3.5}{0.0314}$$

or, $= 1.969$ m/sec. **Ans.**

EXAMPLE 1.10

Example 1.11 A pipe for supplying water having a diameter of 40 cm and connecting with two different sizes of pipes having diameter as 15 cm and 10 cm respectively. Determine the followings

(i) Discharge through 40 cm diameter pipe if the average velocity is 2 m/sec

(ii) Water velocity through 15 cm if the average velocity of water through 10 cm diameter pipe is 1.5 m/sec

Solution

Given data:

Diameter at point 1, D_1=40 cm=0.40 m

$$A_1 = \frac{\pi}{4} D_1^2$$

$$= \frac{\pi}{4} \times 0.4^2$$

or, $$= 0.1256 \text{ m}^2$$

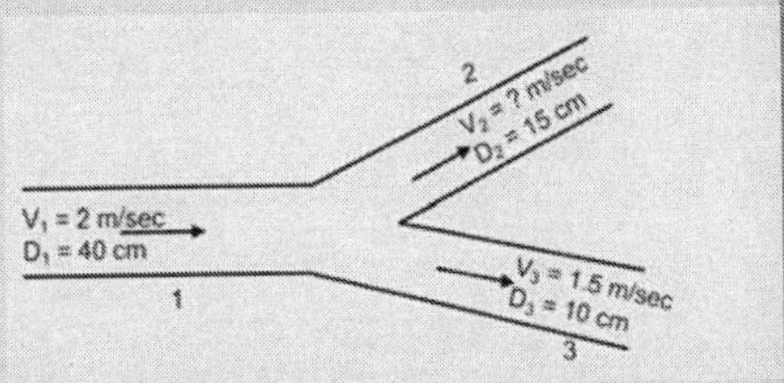

Velocity at point 1, v_1=2 m/s

Diameter at point 2, D_2

$$D_2 = 15 \text{ cm} = 0.15 \text{ m}$$

$$A_2 = \frac{\pi}{4}(0.15)^2$$

EXAMPLE 1.11

or, $= 0.01767\ m^2$

Diameter at point 3, $D_3 = 0.10$ m

$$A_3 = \frac{\pi}{4}(0.10)^2$$

$$A_3 = 0.007853\ m^2$$

Then, according to continuity equation

$$Q_1 = Q_2 + Q_3$$

The discharge Q_1 in pipe 1 is given by

$$Q_1 = A_1 v_1$$

$$= 0.1256 \times 2$$

or, $= 0.2512 \frac{m^3}{s}$. **Ans.**

The discharge Q_2 in pipe 2 is given by

$$Q_2 = A_2 v_2$$

$$= 0.01767 \times V_2$$

The discharge Q_3 in pipe 3 is given by

$$Q_3 = A_3 v_3$$

$$= 0.007853 \times 1.5$$

or, $= 0.01177$

Substituting the values of Q_1 and Q_3, we get

$$0.2512 = Q_2 + 0.01177$$

$$Q_2 = 0.2512 - 0.01177 = 0.23943\ m^3/s$$

The discharge Q_2 in pipe 2 is given by

$$Q_2 = A_2\ v_2 = 0.01767 \times v_2$$

or, $0.23943 = 0.01767 \times v_2$

$$\therefore v_2 = \frac{0.23943}{0.01767} = 13.55 \frac{m}{s}. \textbf{ Ans.}$$

EXAMPLE 1.11

Example 1.12 In an oil refinery, oil is supplied through a pipe having a diameter of 30 cm with a velocity of 2.5 m/s. The quantity of oil is required to be supplied through a pipe of with a diameter of 25 cm. Determine the velocity of the oil through this pipe and total quantity of oil per sec assuming the specific gravity of oil as 0.92.

Solution

Given data:

At section 1

Diameter at point 1, $D_1 = 30 \text{ cm} = 0.3 \text{ m}$

$$A_1 = \frac{\pi}{4} D_1^2$$

or, $= \frac{\pi}{4} \times 0.3^2$

or, $= 0.07068 \text{ m}^2$

Velocity at point 1, $v_1 = 2.5 \text{ m/s}$

At section 2

Diameter at point 2, $D_2 = 25 \text{ cm}$

$$D_2 = 0.25 \text{ m}$$

$$A_2 = \frac{\pi}{4} D_2^2$$

or, $= 0.0490 \text{ m}^2$

Applying continuity equation at sections 1 and 2,

$$A_1 v_1 = A_2 v_2$$

$$0.07068 \times 2.5 = 0.0490 \times v_2$$

$$v_2 = \frac{0.07068 \times 2.5}{0.0490}$$

or, $= 3.606 \text{ m/s}$. **Ans.**

Now, Sp. gr. of oil $= \frac{\rho_{oil}}{\rho_{water}}$

ρ_{oil} = Sp. gr. of oil $\times\ \rho_{water}$

Density of oil $= 0.92 \times 1000$

Density of oil $= 920 \text{ kg/m}^3$

Total quantity of oil is calculated by the given by

Total quantity of oil = Mass density $\times$ discharge $= \rho \times A_1 \times v_1$

Total quantity of oil $= 920 \times 0.07068 \times 2.5 = 162.564$ kg/s. **Ans.**

EXAMPLE 1.12

Example 1.13 A nozzle having inlet and outlet diameters as 30 cm and 15 cm respectively. If the velocity of water at inlet is 3 m/sec, find out the velocity of the water jet at outlet, velocity head at inlet and outlet and discharge

Solution.

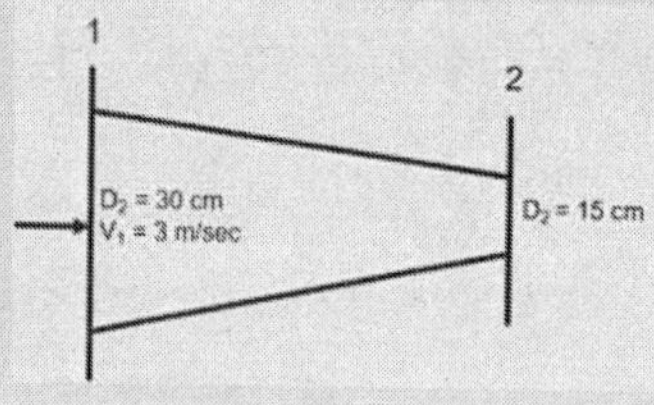

Given data:

Inlet diameter, $D_1 = 30\text{ cm} = 0.3\text{ m}$

$$A_1 = \frac{\pi}{4} D_1^2$$

$$= \frac{\pi}{4}(0.3)^2$$

or, $$= 0.07069\text{ m}^2$$

$v_1 = 3\text{ m/s}$

Outlet diameter, $D_2 = 0.15\text{ m}$

$$A_2 = \frac{\pi}{4}(0.15)^2$$

or, $$= 0.017671\text{ m}^2$$

Velocity head at section 1 is given by

$$\frac{v_1^2}{2g} = \frac{3^2}{2 \times 9.81}$$

$$\frac{v_1^2}{2g} = 0.458\text{ m. } \textbf{Ans.}$$

To find v_2, apply continuity equation at 1 and 2

$$A_1 v_1 = A_2 v_2$$

$$v_2 = \frac{A_1 v_1}{A_2}$$

$$= \frac{0.07069}{0.017671} \times 3.0$$

or, $$= 12.00\frac{\text{m}}{\text{s}}. \textbf{ Ans.}$$

Velocity head at section 2 is given by

$$\frac{v_2^2}{2g} = \frac{12^2}{2 \times 9.81}$$

or, $$= 7.3394\text{ m. } \textbf{Ans.}$$

EXAMPLE 1.3

1.7 MOMENTUM EQUATION

Momentum equation can be derived based on law of conservation of momentum applying Newton's second law of motion.

Let, m is mass of fluid, dv is change in velocity over the given time of dt.

Therefore,

change in momentum $= m.dv$

and rate of change of momentum $= \frac{m.dv}{dt}$

As per the Newton's second law of motion,

Net force (F) in the same direction = rate of change of momentum

$$F = \frac{m.dv}{dt} \tag{1.20}$$

If m is constant, then

$$F = \frac{d(mv)}{dt} \tag{1.21}$$

Eq. (1.21) can be written as

$$F.dt = d(mv) \tag{1.22}$$

Eq. 1. 22 known as Impulse-momentum equation.

1.8 BERNOULLI'S EQUATION

As per the law of conservation of energy, energy contained in a fluid will remain same at different locations of fluid carrying passage i.e., the sum of pressure energy, kinetic energy and potential (elevation) remain constant.

Further, the energy can be expressed in terms of the different heads corresponding to the above-mentioned energy as 'pressure head' $\left(\frac{p}{\rho g}\right)$, 'kinetic head' $\left(\frac{v^2}{2g}\right)$ and 'potential (elevation) head' (z).

The Bernoulli's equation can be expressed in terms of these heads with the assumptions as fluid is real, incompressible and non-viscous. However, it is not possible to have an ideal fluid hence there will be some resistance between fluid layers which will be accounted as head loss (h_L).

Considering two locations of a pipe as shown in Fig. 1.5 the Bernoulli's equation can be derived.

Let

$\frac{p_1}{\rho g}$ and $\frac{p_2}{\rho g}$ are the pressure head at location 1 and 2 respectively.

$\frac{v_1^2}{2g}$ and $\frac{v_2^2}{2g}$ are the velocity head at location 1 and 2 respectively.

z_1 and z_2 are the potential head at location 1 and 2 respectively.

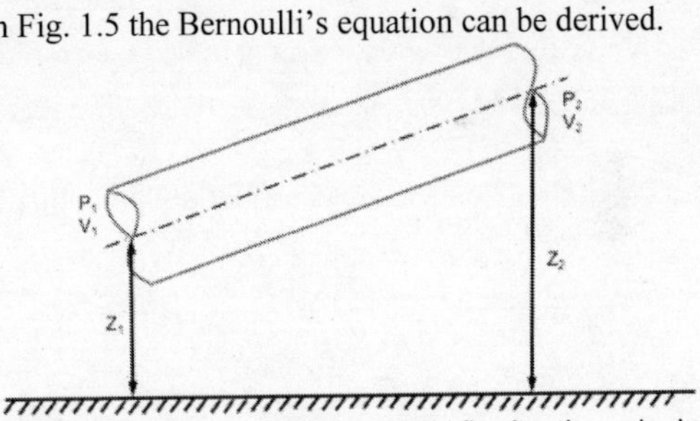

Fig. 1.5: Different locations of fluid flowing through pipe

$$\frac{p_1}{\rho g} + \frac{v_1^2}{2g} + z_1 = \frac{p_2}{\rho g} + \frac{v_2^2}{2g} + z_2 + h_L \tag{1.23}$$

Eq. 1.23 represent the Bernoulli's equation under the assumptions as fluid is real, incompressible and non-viscous.

1.9 PRACTICAL APPLICATIONS OF BERNOULLI'S EQUATION

Bernoulli's equation is considered as an important equation for practical applications and various flow measuring devices are designed based on Bernoulli's equation. Followings are some of the flow measuring devices;

(i) Venturimeter
(ii) Orifice meter
(iii) Pitot-tube

1.9.1 Venturimeter

Flow rate of liquids in pipes can be measured with the help of a venturimeter which is device based on Bernoulli's equation. A venturimeter consists of (i) converging section (ii) throat, and (iii) diverging section as shown in Fig. 1.6.

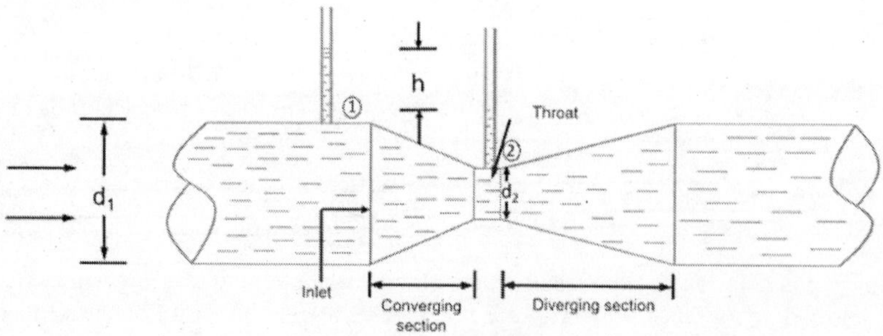

Fig. 1.6: Venturimeter

Consider a venturimeter fitted in a horizontal pipe through which a fluid is flowing.

Let $d_1$is inlet diameter, p_1 is pressure, v_1 is velocity of fluid at section 1 and d_2, p_2, v_2 are corresponding values at section 2.

Assuming friction losses are negligible, the Bernoulli's equation (1.23) at section 1 and 2 can be written as

$$\frac{p_1}{\rho g} + \frac{v_1^2}{2g} + z_1 = \frac{p_2}{\rho g} + \frac{v_2^2}{2g} + z_2 \tag{1.24}$$

The venturi is fitted horizontally, then $z_1 = z_2$

$$\frac{p_1 - p_2}{\rho g} = \frac{v_2^2}{2g} - \frac{v_1^2}{2g} \tag{1.25}$$

The pressure difference between section 1 and 2 can be represented in terms of head as

$$\frac{p_1 - p_2}{\rho g} = h$$

Putting above value in the Eq. 1.25,

$$h = \frac{v_2^2}{2g} - \frac{v_1^2}{2g} \tag{1.26}$$

Further, if the fluid incompressible then continuity equation can be applied between section 1 and 2 and is written as

$$A_1 v_1 = A_2 v_2$$

$$or \qquad v_1 = \frac{A_2 v_2}{A_1} \qquad (1.27)$$

Where, A_1 $(=\frac{\pi}{4}{d_1}^2)$ is cross sectional area at section 1 and A_2 $(=\frac{\pi}{4}{d_2}^2)$ is cross sectional area at section 2

Putting the value of v_1 in Eq. (1.27)

$$h = \frac{v_2^2}{2g} - \frac{(\frac{A_2 v_2}{a_1})^2}{2g}$$

or $\quad v_2^2 = 2gh\frac{a_1^2}{a_1^2 - a_2^2}$

Therefore, $\quad v_2 = \sqrt{2gh\frac{a_1^2}{a_1^2 - a_2^2}} \qquad (1.28)$

$$= \frac{A_1}{\sqrt{A_1^2 - A_2^2}}\sqrt{2gh} \qquad (1.29)$$

SCAN ME

For better understanding venturimeter

Now, flow rate or discharge , $Q = A_2 v_2$

$$= \frac{A_2 A_1}{\sqrt{A_1^2 - A_2^2}} \times \sqrt{2gh} \qquad (1.30)$$

Considering the discharge co-efficient, C_d for the venturimeter, the final equation to measure the actual discharge is expressed as

$$Q_{act} = C_d \times \frac{A_2 A_1}{\sqrt{A_1^2 - A_2^2}} \times \sqrt{2gh} \qquad (1.30)$$

In the above equation, h is the manometric head measured by a differential U-tube manometer and its value can be expressed for different cases as given in Table 1.1.

Table 1.1: Value of manometric head (h) for different cases

S.No.	Cases	Manometric head, h	
		Horizontal manometer	**Inclined manometer**
1	Manometer liquid is heavier than liquid flowing through pipe	$x\left[\frac{S_h}{S_o} - 1\right]$	$\left(\frac{p_1}{\rho g} + z_1\right) - \left(\frac{p_2}{\rho g} + z_2\right) = x\left[\frac{S_h}{S_o} - 1\right]$
2	Manometer liquid is lighter than liquid flowing through pipe	$x\left[1 - \frac{S_l}{S_o}\right]$	$\left(\frac{p_1}{\rho g} + z_1\right) - \left(\frac{p_2}{\rho g} + z_2\right) = x\left[1 - \frac{S_l}{S_o}\right]$

Where, S_h is specific gravity of the heavier liquid, S_l is specific gravity of the lighter liquid, S_o is Specific gravity of the liquid flowing through pipe and x is manometer reading (difference in manometer fluid columns)

1.9.2 Orifice Meter

Flow measurement of liquid in pipe can also be measured by orifice meter. It is a circular plate having a sharp edge circular cut in the center called as orifice. The orifice diameter generally is kept half of the plate diameter. The orifice plate is attached with the pipe through flanges as shown in Fig. 1.7.

The expression to measure the discharge by orifice meter can be derived based on Bernoulli's as follows;

Let, a_1 is cross-sectional area, p_1 is pressure, v_1 is velocity of fluid at section 1 and a_2, p_2, v_2 are corresponding values at section 2.

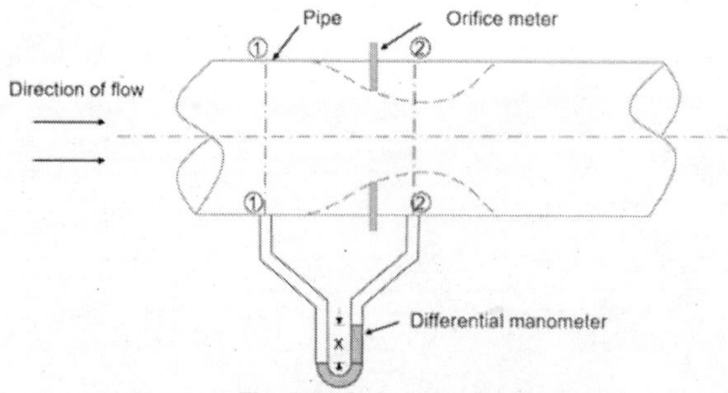

SCAN ME
For better understanding orifice meter

Fig. 1.7: Orifice meter

Based on Bernoulli's equation

$$\frac{p_1}{\rho g} + \frac{v_1^2}{2g} + z_1 = \frac{p}{\rho g} + \frac{v_2^2}{2g} + z_2 \tag{1.31}$$

Or $$\left(\frac{p_1}{\rho g} + z_1\right) - \left(\frac{p_2}{\rho g} + z_2\right) = \frac{v_2^2}{2g} - \frac{v_1^2}{2g} \tag{1.32}$$

In the above $\left(\frac{p_1}{\rho g} + z_1\right) - \left(\frac{p_2}{\rho g} + z_2\right)$ can be expressed as differential head, h

Therefore, $$h = \frac{v_2^2}{2g} - \frac{v_1^2}{2g} \tag{1.33}$$

or $$v_2 = \sqrt{2gh + v_1^2} \tag{1.34}$$

Further, the continuity equation can be applied between section 1 and 2 and is written as

$$Q = A_1 v_1 = A_2 v_2 \tag{1.35}$$

or $$v_1 = \frac{A_2 v_2}{A_1} \tag{1.36}$$

Where, A_2 represents the area at the vena-contracta. If A_o is the area of orifice then, coefficient of contraction C_c is represented as

$$C_c = \frac{A_2}{A_0} \tag{1.37}$$

Putting the value of Cc in Eq. 36

$$v_1 = \frac{A_0 C_c}{A_1} v_2 \tag{1.38}$$

Further, putting the value of v_1 in Eq. 1.34

$$v_2 = \sqrt{2gh + \frac{A_0^2 C_c^2 v_2^2}{a_1^2}} \tag{1.39}$$

$$or \qquad v_2^2 = 2gh + \left(\frac{A_0}{A_1}\right)^2 C_c^2 v_2^2$$

$$or \qquad v_2 = \frac{\sqrt{2gh}}{\sqrt{1-\left(\frac{A_0}{A_1}\right)^2 C_c^2}} \tag{1.40}$$

Therefore, the flow rate or discharge, Q

$$Q = \frac{A_0 C_c \sqrt{2gh}}{\sqrt{1 - \left(\frac{A_0}{A_1}\right)^2 C_c^2}} \tag{1.41}$$

By considering the coefficient of discharge, C_d of orifice plate, the final expression for actual discharge measurement is written as

$$Q_{act} = \frac{A_0 C_d A_1 \sqrt{2gh}}{\sqrt{A_1^2 - A_0^2}} \tag{1.42}$$

1.9.3 Pitot-tube

Pitot-tube is used to measure the point velocity of flow through a pipe, channel or duct. The working principle of pitot tube is based on Bernoulli's equation i.e., when the velocity of flow at a point becomes zero a maximum pressure is observed at that point. A pitot tube is a tube which is bent at 90^0 and placed inside the flow conduit as shown in Fig. 1.8.

In similar lines of other devices, Bernoulli's equation can be applied to derive the expression for fluid discharge measurement. Let p_1 is pressure, v_1 is velocity of fluid at point 1 and P_2, v_2 are corresponding values at point 2, H is depth of tube in the liquid and h is rise of liquid in the tube above the free surface.

Based on Bernoulli's equation between point 1 and 2, then

$$\frac{P_1}{\rho g} + \frac{v_1^2}{2g} + z_1 = \frac{P_2}{\rho g} + \frac{v_2^2}{2g} + z_2 \tag{1.43}$$

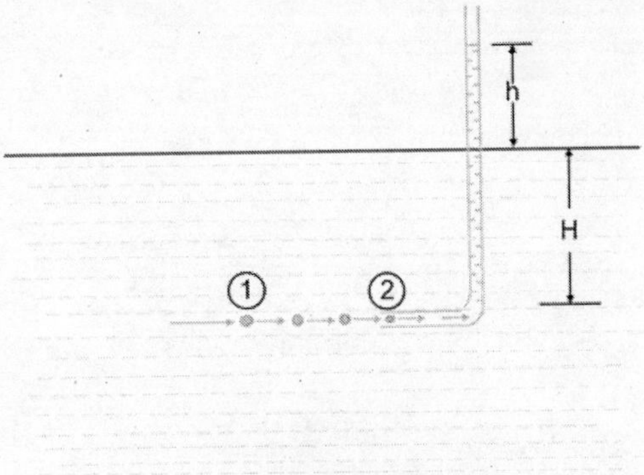

Fig. 1.8: Pitot tube

But $z_1 = z_2$ as points (1) and (2) are on the same line and $v_2 = 0$

$\frac{p_1}{\rho g}$ = pressure head at point 1 = H

$\frac{p_2}{\rho g}$ = pressure head at point 2 = $(h + H)$

SCAN ME

For better understanding pitot-tube

Putting the above values in Eq. 1.43, we get

$$H + \frac{v_1^2}{2g} = (h + H) \tag{1.44}$$

$$h = \frac{v_1^2}{2g}$$

or $$v_1 = \sqrt{2gh} \tag{1.45}$$

v_1 represents the theoretical velocity, however, actual velocity can be expressed as

$$(v_1)_{act} = C_v\sqrt{2gh} \tag{1.46}$$

Where, C_v is Co-efficient of velocity of pitot-tube

Therefore, he final velocity at any point measured by pitot tube is given by

$$v = C_v\sqrt{2gh} \tag{1.47}$$

EXAMPLE 1.14

Example 1.14 Orientation of a pipe having 500 mm diameter as shown in Figure given below. Determine the head losses through the pipe if flow velocity is 10 m/sec. The values of other parameters are as given below

(i) Pressures at point A and B are as 30 N/cm² and 23 N/cm² respectively

(ii) Elevations z_A and z_B are as 25 m and 27 m respectively

Solution:

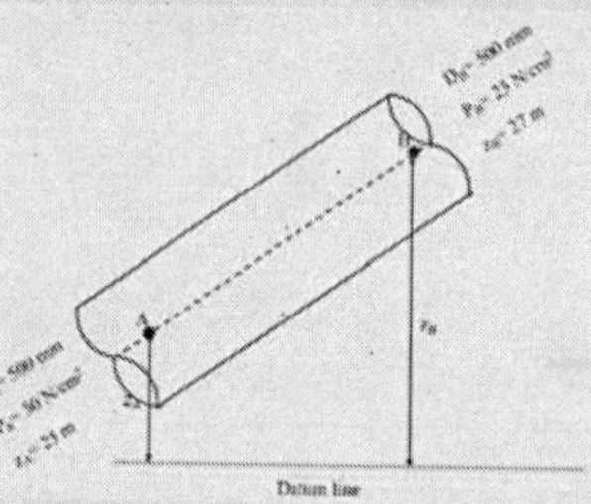

Given data:

Pipe diameter, D = 400 mm = 0.4 m

Velocity at point A, $v_A = 10$ m/s

Pressure at point A,

$$p_A = 30\frac{\text{N}}{\text{cm}^2} = 30 \times 10^4 \text{ N/m}^2$$

Elevation at point A, $z_A = 25$ m

Total energy at point A is given by

$$E_A = \frac{p_A}{\rho g} + \frac{v_A^2}{2g} + z_A$$

$$= \frac{30 \times 10^4}{1000 \times 9.81} + \frac{10^2}{2 \times 9.81} + 25$$

or, $= 30.58 + 5.096 + 25$

or, $= 60.676$ m

Pressure at point B, is given by

$$P_B = 23 \text{ N/cm}^2$$
$$= 23 \times 10^4 \text{ N/m}^2$$

Elevation at point B, $z_B = 27$ m

Velocity at point B, $v_B = v_A = 10$ m/s

Total energy at B,

$$E_B = \frac{p_B}{\rho g} + \frac{v_B^2}{2g} + z_B$$

$$E_B = \frac{23 \times 10^4}{1000 \times 9.81} + \frac{10^2}{2 \times 9.81} + 27$$

$$E_B = 55.542 \text{ m}$$

Loss of energy $= E_A - E_B = 60.676 - 55.542 = 5.134$ m. **Ans.**

EXAMPLE 1.14

Example 1.15 To measure the flow rate of water through a pipe a venturi meter is fitted with pipe horizontally. The specifications of venturimeter are as given blow:

(i) Inlet diameter: 20 cm

(ii) Throat diameter: 10 cm

(iii) Manometer reading for pressure difference across the venturi: 15 cm of mercury

(iv) Coefficient of discharge: 0.98.

Solution:

Given data:

Inlet diameter, $d_1 = 20$ cm

$$A_1 = \frac{\pi}{4} d_1^2$$
$$= \frac{\pi}{4}(20)^2$$
$$= 314.16 \text{ cm}^2$$

EXAMPLE 1.15

EXAMPLE 1.15

Throat diameter, $d_2 = 10$ cm

$$A_2 = \frac{\pi}{4} \times 10^2 = 78.54 \text{ cm}^2$$

Coefficient of discharge, $C_d = 0.98$

Reading of differential manometer: $x = 15$ cm of mercury.

Difference of pressure head is given by

$$h = x\left[\frac{S_h}{S_o} - 1\right]$$

where S_h = Sp. gravity of mercury = 13.6,

S_o = Sp. gravity of water = 1

$$h = 15\left[\frac{13.6}{1} - 1\right] = 15 \times 12.6 \text{ cm} = 189 \text{ cm of water.}$$

The discharge through venturimeter is given by equation

$$Q = C_d \frac{A_1 A_2}{\sqrt{A_1^2 - A_2^2}} \times \sqrt{2gh}$$

$$= 0.98 \times \frac{314.16 \times 78.54}{\sqrt{(314.16)^2 - (78.54)^2}} \times \sqrt{2 \times 9.81 \times 189}$$

$$= \frac{1472476.134}{\sqrt{98696.51 - 6168.53}} = \frac{1472476.134}{304.18}$$

$= 4840.805 \text{ cm}^3/\text{s} = \frac{4840.805}{1000}$ lit/sec = 4.8408 lit/sec. **Ans.**

EXAMPLE 1.6

Example 1.16 Determine the resultant force experienced by a right-angle bend fitted with a pipe having a diameter of 250 mm in horizontal plane. The total discharge flowing through the pipe and bend is 0.250 m³/sec. Consider the pressures at the inlet and outlet of the bend as 24 N/cm² and 23 N/cm², respectively.

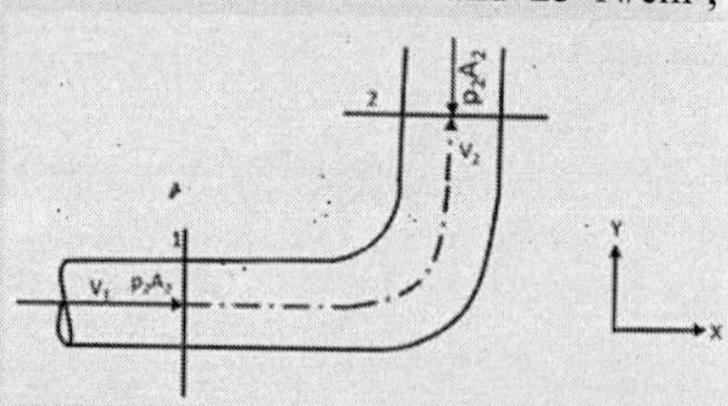

Solution:

Given data:

Density of water, $\rho = 1000 \text{ kg/m}^3$

Pipe diameter, $D = 250$ mm = 0.25 m

$A = A_1 = A_2 = 0.0491 \text{ m}^2$

Total discharge, $Q = 0.25\ \text{m}^3/\text{s}$

$$v = v_1 = v_2 = \frac{Q}{A} = \frac{0.25}{0.0491} = 5.092\ \text{m/s}$$

$$\theta = 90°$$

$$v_{1x} = v_1 = 5.092\ \text{m/s}$$

$$v_{2x} = 0$$

$$v_{1y} = 0$$

$$v_{2y} = v_2 = 5.092\ \text{m/s}$$

Pressure at point 1

$$p_1 = 24\ \text{N/cm}^2 = 24 \times 10^4\ \text{N/m}^2 = 240000\ \text{N/m}^2$$

Pressure at point 2

$$p_2 = 23\ \text{N/cm}^2 = 23 \times 10^4\ \text{N/m}^2 = 230000\ \text{N/m}^2$$

Force on bend along x-axis

$$F_x = \rho Q[v_{1x} - v_{2x}] + (p_1A_1)_x + (p_2A_2)_x$$
$$= 1000 \times 0.25[5.092 - 0] + 240000 \times 0.0491 + 0$$
$$= 1273 + 11784 = 13057\ \text{N}$$

Force on bend along y-axis,

$$F_y = \rho Q[v_{1y} - v_{2y}] + (p_1A_1)_y + (p_2A_2)_y$$
$$= 1000 \times 0.25[0 - 5.092] + (-230000 \times 0.0491)$$
$$= -1273 - 11293 = -12566\ \text{N}$$

∴ Resultant force,

$$F_R = \sqrt{F_x^2 + F_y^2} = \sqrt{(13057)^2 + (-12566)} = 18121.52\ \text{N}.\ \textbf{Ans.}$$

and

$$\tan\theta = \frac{F_y}{F_x} = \frac{12566}{13057} = 0.9624$$

$$\theta = 43°9'.\ \textbf{Ans.}$$

EXAMPLE 1.16

Example 1.17 Determine the force exerted by water on bend as shown in Figure below. The inlet and outlet diameters of the bend are as 0.5 m and 0.25 m respectively. Take the value of pressure intensity at inlet as 7.5 N/cm² and flow rate of 0.5 m³/s.

Solution:

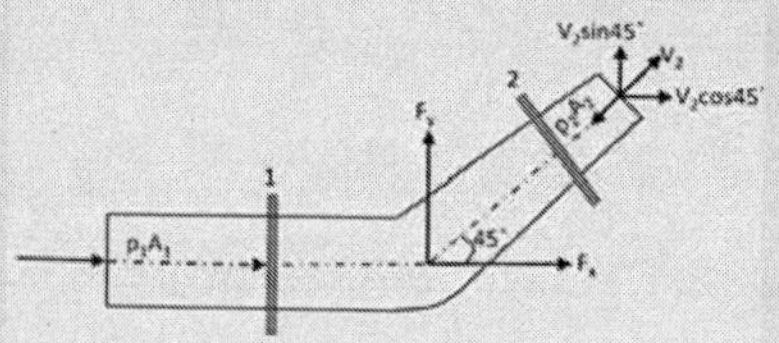

Given data:

Bend angle, $\theta = 45^\circ$

Pipe diameter at point 1, D_1 =0.5 m

$$A_1 = \frac{\pi}{4} \times 0.5^2 = 0.1963 \text{ m}^2$$

Pipe diameter at point 2, D_2 =0.25 m

$$A_2 = \frac{\pi}{4} \times 0.25^2 = 0.0491 \text{ m}^2$$

Pressure at inlet is given by

$$p_1 = 7.5 \text{ N/cm}^2 = 7.5 \times 10^4 \text{ N/m}^2$$

$$Q = 0.5 \text{ m}^3/\text{s}$$

The velocity at point 1 is given by

$$v_1 = \frac{Q}{A_1} = \frac{0.5}{0.1963} = 2.547 \text{ m/s}$$

The velocity at point 2 is given by

$$v_2 = \frac{Q}{A_2} = \frac{0.5}{0.0491} = 10.1833 \text{ m/s}.$$

Applying Bernoulli's equation at sections (1) and (2), we get

$$\frac{p_1}{\rho g} + \frac{v_1^2}{2g} + z_1 = \frac{p_2}{\rho g} + \frac{v_2^2}{2g} + z_2$$

Elevation is same, $z_1 = z_2$

$$\frac{p_1}{\rho g} + \frac{v_1^2}{2g} = \frac{p_2}{\rho g} + \frac{v_2^2}{2g}$$

$$\frac{7.5 \times 10^4}{1000 \times 9.81} + \frac{2.547^2}{2 \times 9.81} = \frac{p_2}{\rho g} + \frac{10.183^2}{2 \times 9.81}$$

$$7.65 + 0.3306 = \frac{p_2}{\rho g} + 5.285$$

$$\frac{p_2}{\rho g} = 7.9806 - 5.285 = 2.6956 \text{ m}$$

Force on bend along x-axis

$$F_x = \rho Q[V_1 - V_2 \cos\theta] + p_1 A_1 - p_2 A_2 \cos\theta$$

$$F_x = 1000 \times 0.5[2.547 - 10.1813\cos 45^\circ] + 7.5 \times 10^4 \times .1963 - 2.64 \times 10^4 \times .0491 \times \cos 45^\circ$$

$$F_x = -2326.13 + 14722.5 - 916.58 = 24959.6 - 5048.2$$

$$F_x = 11479.79 \text{ N}$$

EXAMPLE 1.17

Force on bend along y-axis

$$F_y = \rho Q[-V_2 \sin\theta] - p_2 A_2 \sin\theta$$

$$F_y = 1000 \times 0.5[-10.1813 \sin 45°] - 2.64 \times 10^4 \times .0491 \times \sin 45°$$

$$F_y = -3599.63 - 916.58 = -4516.21\text{N}$$

Resultant force, is given by

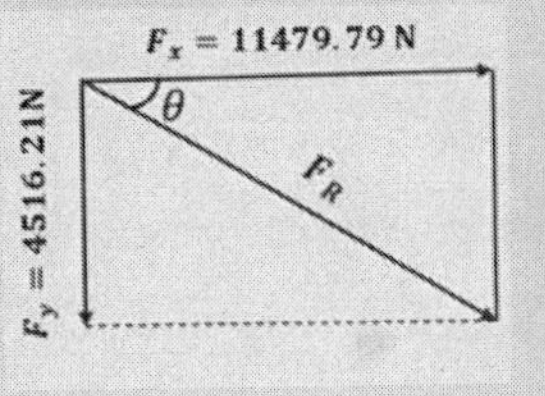

$$F_R = \sqrt{F_x^2 + F_y^2}$$

$$F_R = \sqrt{(11479.79)^2 + (-4516.21)^2}$$

$F_R = 12336.196$ N. **Ans.**

The angle made by resultant force with x-axis is given by

$$\tan\theta = \frac{F_y}{F_x} = \frac{4516.21}{11479.79} = 0.3934$$

$$\therefore \theta = \tan^{-1} .3934 = 21°47'. \textbf{ Ans.}$$

EXAMPLE 1.17

Example 1.18 An orifice meter is fitted in a pipe of 10 cm diameter to measure the discharge of a fluid. The specifications of the orifice meter areas; orifice diameter: 5 cm, pressure gauge readings at upstream and downstream readings: 16 N/cm² and 8 N/cm² and coefficient of discharge: 0.55.

Solution.

Given data:

Diameter of orifice, $d_0 = 5$ cm

Area of orifice, $A_0 = \frac{\pi}{4}(5)^2$

$$A_0 = 19.634 \text{ cm}^2$$

Diameter of pipe, $d_1 = 10$ cm

Area of pipe, $A_1 = \frac{\pi}{4}(10)^2$

A_1=78.539 cm²

Coefficient of discharge, $C_d = 0.55$

Pressure at point 1, $p_1 = 16 \text{ N/cm}^2 = 16 \times 10^4 \text{ N/m}^2$

Pressure head at point 1

$$\frac{p_1}{\rho g} = \frac{16 \times 10^4}{1000 \times 9.81}$$

EXAMPLE 1.18

EXAMPLE 1.18

$= 16.309$ m of water

Pressure at point 2, $p_2 = 8\ \mathrm{N/cm^2} = 8 \times 10^4\ \mathrm{N/m^2}$

Pressure head at point 2

$$\frac{p_2}{\rho g} = \frac{8 \times 10^4}{1000 \times 9.81}$$

$= 8.154$ m of water

Difference in pressure head is given by

$$h = \frac{p_1}{\rho g} - \frac{p_2}{\rho g}$$

$$h = 16.309 - 8.154$$

$h = 8.155$ m of water $= 815.5$ cm of water

The discharge, Q is given by

$$Q = C_d \frac{A_0 A_1}{\sqrt{A_1^2 - A_0^2}} \times \sqrt{2gh}$$

$$Q = 0.55 \times \frac{19.634 \times 78.539}{\sqrt{(78.539)^2 - (19.634)^2}} \times \sqrt{2 \times 981 \times 815.5}$$

$Q = 14107.375\ \mathrm{cm^3/s} = 14.107$ litres / s. **Ans.**

EXAMPLE 1.19

Example 1.19 Find out the rate of flow of an oil (S = 0.91) in a pipe with the help of an orifice meter whose diameter is 10cm. The pipe diameter is 20 cm and pressure difference across the orifice meter is measured as 35 cm of mercury. Take the value of discharge coefficient as 0.60.

Solution

Given data:

Diameter of orifice, $d_0 = 10$ cm

Area of orifice, $A_0 = \frac{\pi}{4}(10)^2 = 78.539\ \mathrm{cm^2}$

Diameter of pipe, $d_1 = 20$ cm

Area of pipe, $A_1 = \frac{\pi}{4}(20)^2 = 314.16\ \mathrm{cm^2}$

Coefficient of discharge, $C_d = 0.60$

Specific gravity, $S = 0.91$

Reading of differential manometer, $x = 35$ cm of mercury

$\therefore$ Differential head, $h = x\left[\frac{S_g}{S_o} - 1\right] = 35\left[\frac{13.6}{0.91} - 1\right]$ cm of oil

$$h = 35 \times 13.94 = 488.077 \text{ cm of oil}$$

The rate of the flow, Q is given by

$$Q = C_d \frac{A_0 A_1}{\sqrt{A_1^2 - A_0^2}} \sqrt{2gh}$$

$$Q = 0.60 \times \frac{78.539 \times 314.16}{\sqrt{(314.16)^2 - (78.539)^2}} \times \sqrt{2 \times 981 \times 488.077}$$

$Q = 47626.01 \text{ cm}^3/\text{s} = 47.626$ litres/s. **Ans.**

EXAMPLE 1.19

Example 1.20 A pitot tube is fitted in a pipe to measure the flow velocity of a fluid (S=0.8), what will be the flow velocity if the manometer reading fitted to the Pitot-tube tapping is measured as 80 mm of mercury.

Solution.

Given data:

Differential of mercury level, $x = 80$ mm $= 0.08$ m

Specific gravity of oil, $S_o = 0.8$

Specific gravity of mercury, $S_m = 13.6$

Assume, $C_v = 0.98$

Diff. of pressure head,

$$h = x\left[\frac{S_m}{S_o} - 1\right] = 0.08\left[\frac{13.6}{0.8} - 1\right] = 1.28 \text{ m of oil}$$

Velocity of flow is given by

$$= C_v\sqrt{2gh} = 0.98\sqrt{2 \times 9.81 \times 1.28} = 4.91 \text{ m/s. } \textbf{Ans.}$$

EXAMPLE 1.20

UNIT SUMMARY

- Fluid Mechanics is a science which deals with kinematics, dynamics, and statics of fluid motion.
- A fluid may be defined as a substance which is capable of flowing and has no definite shape.
- The mass density (ρ) of a fluid can be defined as the ratio of the mass of a fluid to its volume. It is expressed mathematically as

$$\text{Mass density}, \rho = \frac{\text{Mass of fluid, } m}{\text{Volume of fluid, } V}$$

- Specific volume of a fluid is the volume of fluid per unit mass which is reciprocal of mass density, and it is expressed as

$$\text{Specific volume} = \frac{\text{volume of fluid, } V}{\text{mass of fluid, , } m}$$

- Specific weight (γ) is defined as the ratio of weight of fluid and its volume occupied.

$$\gamma = \frac{\text{mg}}{\text{V}} = \rho g$$

- Specific gravity (S) is a dimensionless quantity; it is expressed in terms of weight density of water in case of liquids and in terms of weight density of air for gases.
  It can be expressed as

$$S \text{ (for liquids)} = \frac{\text{Weight density (density) of liquid}}{\text{Weight density (density) of water}}$$

- As per the Newton's law of viscosity, shear stress is proportional to the rate of change of velocity with respect to y. It can be expressed as;

$$\tau = \mu \frac{dv}{dy}$$

  The unit of dynamic viscosity is poise or Ns/m^2 (1 poise=10^{-1} Ns/m^2)
- Kinematic viscosity is defined as the ratio of dynamic viscosity to the density of fluid and expressed as;

$$\nu = \frac{\text{Viscosity } (\mu)}{\text{Density } (\rho)}$$

  The unit of kinematic viscosity is m^2/sec
- Compressibility of fluid is the measure of elasticity, that is represented by bulk modulus of elasticity, K, which is the ratio of compressive stress to volumetric strain.

$$\text{Bulk modulus, K} = \frac{\text{Increase of pressure (compressive stress)}}{\text{Volumetric strain}}$$

and

$$\text{Compressibility} = \frac{1}{\text{Bulk modulus, K}}$$

- Surface tension is the property of the liquid surface film to exert a tensile force and it is denoted by σ. The pressure intensity(p) due to surface tension three different cases as
  (i) pressure intensity inside a droplet
  $$p = \frac{2\sigma}{R}$$
  (ii) pressure intensity inside a soap bubble
  $$p = \frac{4\sigma}{R}$$
  (iii) pressure intensity inside liquid jet is considered.
  $$p = \frac{\sigma}{R}$$
- Continuity equation is based on law of conservation of mass, it can be stated that the total mass of fluid flowing through a passage will be same at different cross sections of the passage if there are no energy losses through the passage.
- The continuity equation for an ideal fluid, steady flow and compressible flow expressed by
  $$\rho_1 A_1 v_1 = \rho_2 A_2 v_2$$
- Momentum equation can be derived based on law of conservation of momentum applying Newton's second law of motion and is expressed by
  $$F = \frac{m.dv}{dt}$$
- The Bernoulli's equation can be expressed in terms of pressure head, kinetic head and potential head with the assumptions as fluid is real, incompressible and non-viscous and is expressed as
  $$\frac{P_1}{\rho g} + \frac{v_1^2}{2g} + z_1 = \frac{P_2}{\rho g} + \frac{v_2^2}{2g} + z_2 + h_L$$

Where h_L is head loss due to friction in pipe

- Bernoulli's equation is considered as an important equation for practical applications and various flow measuring devices are designed based on Bernoulli's equation. Followings are some of the flow measuring devices.

(i) Venturimeter

Venturimeter is discharge measuring device and discharge expressed as

$$Q_{act} = C_d \times \frac{A_2 A_1}{\sqrt{a_1^2 - a_2^2}} \times \sqrt{2gh}$$

Where, C_d is the coefficient of discharge

(ii) Orifice meter

Orifice meter is discharge measuring device and discharge expressed as

$$Q = \frac{A_0 C_c \sqrt{2gh}}{\sqrt{1 - \left(\frac{a_0}{a_1}\right)^2 C_c^2}}$$

Where, C_c is the coefficient of contraction

(iii) Pitot-tube

Pitot tube is flow velocity measuring device and velocity expressed as

$$v = C_v \sqrt{2gh}$$

Where, C_v is the coefficient of velocity

EXERCISES

Multiple Choice Questions

1.1. Kinematic viscosity is measured in units

(a) cm^2 / sec (b) gm/cm2-sec

(c) dyne-sec/cm2 (d) gm/cm-sec2

1.2. Fluid specific weight is determined by the following factors:

(a) mass density of the fluid (b) gravitational acceleration

(c) both a. and b. (d) none of the above

1.3. The dynamic viscosity in C.G.S units is expressed as:

(a) Stoke (b) Pa.s

(c) Poise (d) None of the above

1.4. What is the equivalent unit of dynamic viscosity referred to as 'Poise'?

(a) gm-cm/seo2 (b) dyne/cm^2

(c) dyne-sec/cm^2 (d) gm-sec/cm^2

1.5. A fluid has a kinematic viscosity of 0.1 Stokes. What will be the value is m^2/s?

(a) 10^{-5} (b) 10^{-4}

(c) 10^{-3} (d) 10^{-2}

1.6. Liquids rise in capillaries due to

(a) Surface Tension (b) Diffusion

(c) Osmosis (d) Viscosity

1.7. Which of the following statements is true regarding the Bulk Modulus of Elasticity?

(a) The ratio of tensile stress to linear strain

(b) The ratio of tensile stress to volumetric strain

(c) The ratio of compressive stress to linear strain

(d) The ratio of compressive stress to volumetric strain

1.8. The continuity equation can be expressed as follows:

(a) $A_1 v_1 = A_2 v_2$ (b) $\rho_1 A_1 = \rho_2 A_2$

(c) $\rho_1 A_1 v_1 = \rho_2 A_2 v_2$ (d) None of these

1.9. If a liquid enters a pipe of diameter d at velocity v, what will be its exit velocity when the diameter is reduced to 0.5d?

(a) $4v$ (b) $2v$

(c) $0.5v$ (d) v

1.10. How much area is required for a pipeline to transport 100 m^3/s of water at a velocity of 0.25 m/s?

(a) 200 m^2 (b) 300 m^2

(c) 400 m^2 (d) 100 m^2

1.11. Application of Bernoulli's equation

(a) Pitot tube meter (b) Orifice meter

(c) Venturimeter (d) All of the above

1.12. In the case of incompressible fluid flow, what is the effect of reducing the area on the velocity?

(a) first decreases then increases (b) first increases then decreases

(c) decreases (d) increases

1.13. Equation of continuity is based on the principle of conservation of

(a) energy (b) momentum

(c) mass (d) none of the above

1.14. Bernoulli's theorem relates to

(a) Fluid (b) Space

(c) Electricity (d) Light

1.15. The size of a venturimeter is determined by

(a) both pipe diameter and throat diameter (b) angle of diverging section

(c) throat diameter (d) pipe diameter

1.16. To apply Bernoulli's theorem, it is necessary for the fluids

(a) to be at atmospheric pressure (b) must be incompressible

(c) to be of unit density (d) should have high viscosity

1.17. Under what conditions can the Bernoulli equation not be used?

(a) laminar flow (b) steady flow
(c) incompressible fluid (d) Viscous flow

1.18. Which of the following statements is true in regards to the proof of Bernoulli's theorem?
(a) Loss of energy is zero (b) Liquid is viscous
(c) Liquid is taken compressible (d) All of the above

1.19. What is the physical principle behind momentum equation?
(a) First law of thermodynamics (b) Zeroth law of thermodynamics
(c) Newton's first law of motion (d) Newton's second law of motion

1.20. In the momentum equation, the source term is ________
(a) Acceleration (b) Viscous force
(c) Body forces (d) Pressure force

Answers of Multiple-Choice Questions

1.1 (a), 1.2 (c), 1.3 (c), 1.4 (c), 1.5 (a), 1.6 (a), 1.7 (d), 1.8 (c), 1.9 (a), 1.10 (c), 1.11 (d), 1.12 (d), 1.13 (c), 1.14 (a), 1.15 (d), 1.16 (b), 1.17 (d), 1.18 (a), 1.19 (d), 1.20 (c).

Short and Long Answer Type Questions

1.1. What do you understand by the fluid properties of density, weight density, specific volume, and specific gravity?
1.2. Define dynamic viscosity and kinematic viscosity with their units.
1.3. Discuss Newton's law of viscosity.
1.4. What will be the value of dynamic viscosity in poise corresponds to 1 kg/s-m.
1.5. Derive the expression of continuity equation for a 3-dimensional fluid flow.
1.6. What is Bernoulli's equation?
1.7. What are the three different ways to express the Bernoulli equation?
1.8. State the assumptions to derive the Bernoulli equation.
1.9. Define a venturimeter and derive the equation of venturimeter to measure the discharge through a pipe.
1.10. What are the advantages and disadvantages of Orifice meter over a venturimeter?
1.11. Discuss the pitot tube with respect to pitot-static tube.
1.12. Explain how a pitot-tube is used to measure the point velocity in a fluid.
1.13. With a neat sketch show all the components of a pitot tube, venturimeter and orifice meter.
1.14. Discuss the operation of pitot tube, Venturimeter and Orifice meter to measure flow rate of a fluid.
1.15. Derive the momentum equation and discuss the application of it to determine the force considering the practical system.

Numerical Problems

1.16. Determine the value of specific weight, mass density and specific gravity of an oil having a volume of 5 m3 with a weight of 5000 kg(f).

1.17. What will be the dynamic viscosity of a fluid flowing between the plates placed at a distance of 0.03 mm if one plate moves with a velocity of 0.5 m/sec over other fixed plate. The force is exerted between the plates is as 0.15 kg(f)/m^2 to maintain the given value of speed.

1.18. Determine the terminal velocity which required to be attained by a cubical body having its edge as 20 cm and weight of 18 kg(f) which is sliding down over an inclined plane at 18° to the horizontal. Consider the fill thickness as 0.02 mm for a thin film of oil having viscosity 0.022 kg(f)-s/cm^2.

1.19. Find out diameter of a water droplet assuming surface tension between air and water as 0.08 N/m at 20°C, if the difference between internal and external pressure is 0.020 N/cm^2.

1.20. If the diameters of small and large piston used in hydraulic jack are 4 cm and 12 cm respectively, determine load lift by the large piston against a force of 100 N for the cases (i) level of both the pistons are same (ii) The large piston is below 40 cm to the small piston. Consider density of the liquid used in the jack is 998 kg/m^3.

1.21. Determine the value of bulk modulus of elasticity of a liquid when it is compressed in a cylinder for zero change in volume under a pressure increase from 100 N/cm^2 to 200 N/cm^2.

1.22. For a isothermal process determine the pressure inside a cylinder contains volume of air as 1 m3 for absolute pressure as 40 N/cm at 0°C. What will be the pressure and temperature in the same case considering the k value as 1.4 for air.

1.23. Determine the shear stress for a shaft having a diameter of 130 mm and rotating with 250 r.p.m in a journal bearing. If an oil having viscosity of 1 poise is used to lubricate the bearing with a clearance as 1.55 mm between shaft and journal bearing.

1.24. How much force should be applied to a plunger of a hydraulic press having a diameter of 3 cm fitted with a cylinder of 200 cm diameter to lift a weight of 25000 N.

1.25. Considering the atmospheric pressure equivalent to 750 mm of mercury whose specific gravity is 13.6, determine the absolute pressure and gauge pressure at a point below 5 m under a liquid (ρ= 1.53 x 103 kg/m^3). Take water density as 1000 kg/m^3.

1.26. A fluid is flowing through an expansion pipe having inlet and outlet diameters as 100 mm and 150 mm respectively. If the velocity of fluid is 3 m/sec at inlet, then find out the velocity at outlet of the pipe.

1.27. In an oil refinery, oil is supplied through a pipe having a diameter of 50 cm with a velocity of 1.5. The quantity of oil is required to be supplied through a pipe of with a diameter of 30 cm. Determine the velocity of the oil through this pipe and total quantity of oil per sec assuming the specific gravity of oil as 0.85.

1.28. For an inclined pipe having 200 mm diameter, determine the head losses through the pipe if flow velocity is 8 m/sec. The height of the upper and lower ends from datum line are 20m and 25m respectively and corresponding pressure values are 25 N/cm^2 and 20 N/cm^2.

1.29. For the given parameters of a venturimeter find out the flow rate of water (i) inlet diameter: 20 cm (ii) Throat diameter: 10 cm (iii) Manometer reading for pressure difference across the venturi: 15 cm of mercury and (iv) Coefficient of discharge: 0.98.

1.30. Determine the resultant force experienced by a right-angle bend fitted with a pipe having a diameter of 200 mm in horizontal plane. The total discharge flowing through the pipe and bend is 0.30 m^3/sec. Consider the pressures at the inlet and outlet of the bend as 23 N/cm^2 and 22 N/cm^2, respectively.

1.31. An orifice meter is fitted in a pipe of 8 cm diameter to measure the discharge of a fluid. The specifications of the orifice meter areas; orifice diameter: 4 cm, pressure gauge readings at upstream and downstream readings: 12 N/cm^2 and 6 N/cm^2 and coefficient of discharge: 0.6.

1.32. Find out the rate of flow of an oil (S = 0.92) in a pipe with the help of an orifice meter whose diameter is 6 cm. The pipe diameter is 12 cm and pressure difference across the orifice meter is measured as 25 cm of mercury. Take the value of discharge coefficient as 0.60.

1.33. A pitot tube is fitted in a pipe to measure the flow velocity of a fluid (S=0.82), what will be the flow velocity if the manometer reading fitted to the pitot-tube tapping is measured as 75 mm of mercury.

REFERENCES AND SUGGESTED READINGS

List of some of the books is given below which may be used for further learning of the subject:

1. Yunus A. Çengel, Fluid Mechanics: Fundamentals and Applications, McGRAW-HILL publication, 2006.
2. P. N. Modi, Hydraulics & Fluid Mechanics including Hydraulics Machines, Rajsons Publications Pvt. Ltd., 2014.
3. R. K. Bansal., "A Textbook of Fluid Mechanics and Hydraulic Machines", Laxmi Publications Pvt. Ltd. 2010.
4. R.K. Rajput., "Fluid Mechanics and Hydraulic Machines", S. Chand & Company Ltd. 2011.
5. S. K. SOM and G Biswas., "Introduction to Fluid Mechanics and Fluid Machines", Tata McGraw-Hill Publishing Company Limited New Delhi 2008.
6. Yunus A. Çengel and John M. Cimbala., "Fluid Mechanics: Fundamentals and Applications", McGraw Hill Publishing. 2006.

2 Fluid Dynamics

UNIT SPECIFICS

Through this unit we have discussed the following aspects:

- *Fluid flow through channels and ducts;*
- *Velocity distribution in open channel;*
- *Concepts of Couette and Poisuielle flow;*
- *Laminar flow through circular conduits and annuli;*
- *Concept of boundary layer and measures of its thickness;*
- *Derive Darcy Weisbach equation and friction factor;*
- *Moody's diagram and problems related to fluid dynamics.*

The practical applications of the topics are discussed for generating further curiosity and creativity as well as improving problem solving capacity in Fluid dynamics.

Besides giving a large number of multiple choice questions as well as questions of short and long answer types marked in two categories following lower and higher order of Bloom's taxonomy, assignments through a number of numerical problems, a list of references and suggested readings are given in the unit so that one can go through them for practice. It is important to note that for getting more information on various topics of interest some QR codes have been provided in different sections which can be scanned for relevant supportive knowledge.

RATIONALE

This fundamental unit on fluid dynamics helps students to get a primary idea about the movement of liquids and gases and forces acting upon them in nature. It explains the forces acting on fluid when particle in motion. All these basic aspects are relevant to study the fluid dynamics properly. Based on Navier-Stokes equation, the unit explains clearly fluid dynamics and its completeness in describing particles motion and also explains their form invariance. All these are discussed at length to develop the subject. Some related problems are given with the solutions for laminar flow between two parallel plates, conduits, annuli and boundary layer, and friction factor which can help further for getting a clear idea of the concern topics on fluid dynamics.

Fluid dynamics is the sub-discipline of fluid mechanics which is an important branch of physical science that deals with forces and energy and their effect when fluid is in motion. Dynamics started its journey by quantifying fluid flow and then explaining it in terms of forces, energy and momentum. This permits one to calculate the forces acting upon the fluid. But at the same time it covers the practical problems in real world and which are related to the design, construction and operation of different types of devices and instruments.

PRE-REQUISITES

Mathematics: Vectors, Calculus (Class XII)

Physics: Mechanics (Class XII)

UNIT OUTCOMES

List of outcomes of this unit is as follows:

U2-O1: Describe fluid flow through channels and ducts

U2-O2: Describe the concepts of Couette flow and Poisuielle flow

U2-O3: Define laminar flow through circular conduits and annuli

U2-O4: Explain the concept of boundary layer and derive the Darcy Weisbach equation

U2-O5: Explain the friction factor and Moody's diagram and solve complex problems

Unit-2 Outcomes	EXPECTED MAPPING WITH COURSE OUTCOMES (1- Weak Correlation; 2- Medium correlation; 3- Strong Correlation)					
	CO-1	CO-2	CO-3	CO-4	CO-5	CO-6
U2-O1	3	3	3	-	3	1
U2-O2	1	1	2	2	1	-
U2-O3	2	1	3	1	2	1
U2-O4	-	-	3	1	2	2
U2-O5	3	3	3	-	3	1

2.1 FLUID FLOW THROUGH CHANNELS AND DUCTS

In practise, if flow of fluids occurs through some passage having cross-section other than circular, the passage will be known as channel or duct. If the fluid is liquid and has free surface flowing under atmospheric pressure, then the channel will be termed as open channel. In case of air or gas flow through a channel where fluid does not have free surface and remains under pressure then channel can be termed as closed channel or duct. The flow of liquids take place in open channel under gravity force and may be classified as (i) natural channel flow and (ii) artificial channel flow. Natural channels usually have irregular cross-section and the roughness of the bed and sides also varies considerably from smooth to very rough like flow in streams, and the rivers or artificial channels built for navigation, irrigation or power generation. On the other hand, the artificial channels have fairly well-defined property as cross-section, bed slope and roughness.

In order to have the exact solutions of fluid flow in open channel, the types of flow are required to be discussed as

(i) steady and unsteady flow

In case of steady flow, the velocity of flow (v) and corresponding discharge (Q) remains constant with respect to time (t) at a particular point (y) in the channel. While, these parameters will vary with time in case of unsteady flow. Mathematically, these can be expressed as;

$dv/dt = 0; dQ/dt = 0; dy/dt = 0$ for steady flow

and

$dv/dt \neq 0; dQ/dt \neq 0; dy/dt \neq 0$ for unsteady flow

(ii) uniform and non-uniform flow

A flow can be said as uniform flow if the velocity of flow (v) and corresponding discharge (Q) remains constant with respect to space (s) for a given time (t). However, these parameters will vary with space for a given time in case of non-uniform flow. Mathematically, these can be expressed as;

$dv/ds = 0; dQ/ds = 0;$ for uniform flow

and

$dv/ds \neq 0; dQ/ds \neq 0;$ for non-uniform flow

(iii) laminar and turbulent flow

Under laminar flow, the fluid particles follow smooth path in layers and each layer moves smoothly over the adjacent layers and do not cross each other. However, in case of turbulent flow the fluid particles follow irregular pattern and fluid layers are inter-mix to each other. The flow can be classified based on Reynolds number, ($Re = \rho v D/\mu$) as;

$Re \leq 600$ for laminar flow and

$Re \geq 2000$ for turbulent flow

$600 > Re < 2000$ for transition state in general.

Where, Re is Reynolds number, D is hydraulic diameter ($=4A_c/P$), A_c is cross-sectional area, P is perimeter, ρ is density of fluid, V is velocity of flow, μ is viscosity of fluid.

(iv) Critical, Sub-Critical and Super - Critical Flow

These terms are related to the specific energy of fluid and critical depth of channel and classified based on Froude number. Critical depth is the depth of maximum discharge, when the specific energy is kept constant. Froude number is an important dimensionless parameter used to describe the flow in open-channel. It can be defined as the ratio of inertia forces to gravity forces and expressed as;

$$Fr = \frac{v}{\sqrt{gh_d}} \tag{2.1}$$

Where, v is the mean velocity of fluid flow, h_d is the hydraulic depth of channel (= wetted area/ width of channel)

$Fr < 1$ for sub-critical which can also be called as tranquil or streaming flow

$Fr = 1$ for critical

$Fr > 1$ for super critical

Open channels are used for liquid flow under atmospheric pressure due to gravity. The liquid may be in uniform or non-uniform, in case of non-uniform the flow is not constant across the cross-section of the channel. The non-uniformity of velocity may be due to presence of the frictional resistance which depends on the shape (cross-section), roughness, alignment and bed slope of channel. For practical applications, the open channels can be designed in different cross-sections such as (i) Rectangular Channel, (ii) Trapezoidal Channel, and (iii) Circular Channel. The semi-circular open channel allows to maximum discharge with frictional resistance are considered efficient cross-section of open channel. However, mostly rectangular and trapezoidal channel sections are recommended.

The cross-sectional area and slope of open channels are considered important parameters to design a channel for a given discharge. A relationship known as Manning equation as expressed by Eq. 2.2 is mostly used to design an open channel.

$$Q = \frac{1}{n} A R_H^{2/3} S_0^{1/2} \tag{2.2}$$

Where, Q is discharge, A is cross-sectional area of channel, n is Manning coefficient, R_H is hydraulic radius and S_0 is bottom slope of the channel.

2.2 POISUIELLE FLOW

The concept of Poisuielle flow deals with the fluid dynamics and it can be defined as the laminar flow between two fixed parallel plates. The condition for Poisuielle flow can be derived by considering two parallel fixed plates placed at a distance of D as shown in Fig. 2.1.

Assuming a fluid element between two fixed plates of a steady, incompressible, laminar flow moving in x direction between two infinite parallel horizontal plates as shown in the figure. It is therefore, the velocity in other two directions will be zero and for steady state continuity equation can be expressed as

$$\frac{\partial v}{\partial x} = 0 \tag{2.3}$$

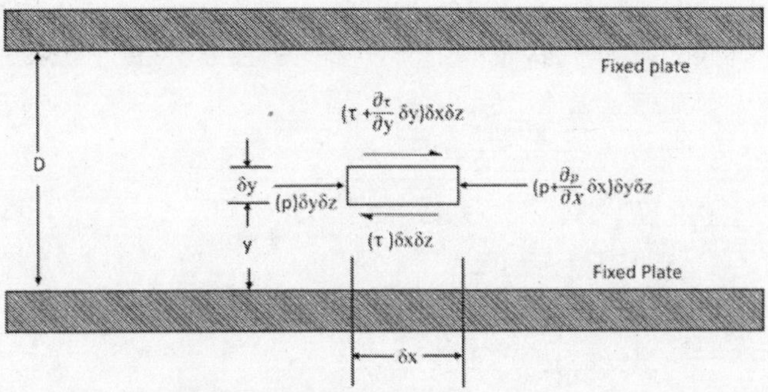

Fig. 2.1: Concept of Poisuielle flow

Based on the forces acting over the fluid element under the considered conditions,

$$(p)\delta y\delta z - \left(p + \frac{\partial p}{\partial x}\delta x\right)\delta y\delta z + \left(\tau + \frac{\partial \tau}{\partial x}\delta y\right)\delta x\delta z - (\tau)\,\delta x\delta z = 0$$

Where, $\tau = \mu \frac{\partial v}{\partial y}$

$$\frac{\partial p}{\partial x} = \mu \frac{d^2 v}{dy^2} \tag{2.4}$$

Considering a pressure drop between the plate over a fluid element,

By integration Eq. 2.4

$$\frac{\partial v}{\partial y} = -\frac{1}{\mu}\frac{\partial p}{\partial x} y + C_1 \tag{2.5}$$

Further Integration of Eq. 2.4

$$v = \frac{1}{2\mu}\frac{\partial p}{\partial x} y^2 + C_1 y + C_2 \tag{2.6}$$

The above Eq. 2.6 represents for laminar flow between two fixed parallel plates or Poisuielle flow.

C_1 and C_2 can be determined using boundary conditions. Since there is a no slip condition at the two boundary surfaces of plates,

$v = 0$ at $y = 0$

$\therefore \quad C_2 = 0$

and $\quad v = 0$ at $y = D$

$\therefore \quad C_1 = -\frac{D}{2\mu}\left(\frac{\partial p}{\partial x}\right)$

Putting the values of C_1 and C_2 in Eq. 2.6, we get

$$v = \frac{1}{2\mu}\left(-\frac{\partial p}{\partial x}\right)(Dy - y^2) \tag{2.7}$$

Maximum velocity, v_{max} occurs at the mid-point between the plates, and can be evaluated by putting $y = D/2$ in Eq. 2.7. Thus,

$$v_{max} = \frac{D^2}{8\mu}\left(-\frac{\partial p}{\partial x}\right) \quad (2.8)$$

The discharge q per unit width can be obtained by integrating dq passing through small strip dy

Since $\quad dq = vdy$

$\therefore \quad q = \int_0^D vdy = \int_0^D \frac{1}{2\mu}\left(-\frac{\partial p}{\partial x}\right)(Dy - y^2)\,dy$

$$q = \frac{D^3}{12\mu}\left(-\frac{\partial p}{\partial x}\right) \quad (2.9)$$

Mean velocity, V is given by

$$V = \frac{q}{D} = \frac{D^2}{12\mu}\left(-\frac{\partial p}{\partial x}\right) \quad (2.10)$$

On comparing Eq. 2.8 and Eq. 2.10, we get

$$V = \frac{2}{3}v_{max} \quad (2.11)$$

From Eq. 2.10, we can get the pressure gradient as

$$\left(-\frac{\partial p}{\partial x}\right) = \frac{12\mu V}{D^2} \quad (2.12)$$

Pressure drop $(p_1 - p_2)$ in the direction of flow is given by

$$(p_1 - p_2) = \int_{p_1}^{p_2} (-\partial p) = \int_{x_1}^{x_2} \frac{12\mu V}{D^2}(\partial x)$$

$$(p_1 - p_2) = \frac{12\mu VL}{D^2} \quad (2.13)$$

Pressure head drop, h_f can be obtained by

$$h_f = \frac{p_1 - p_2}{w} = \frac{12\mu VL}{wD^2} \quad (2.14)$$

Where, w is the specific weight of the fluid

It is observed that the head loss for this flow varies linearly with velocity.

Now, putting Eq. 2.7 into Newton's law of velocity, we get

$$\tau = \mu\frac{\partial}{\partial y}\left[\frac{1}{2\mu}\left(-\frac{\partial p}{\partial x}\right)(Dy - y^2)\right]$$

$$\tau = \left(-\frac{\partial p}{\partial x}\right)\left(\frac{D}{2} - y\right) \quad (2.15)$$

Shear stress varies linearly with the distance from the boundary. It is maximum at the boundary ($y = 0$ or $y = D$) and zero at the centre line ($y = \frac{D}{2}$). Hence,

$$\tau_{max} = \left(-\frac{\partial p}{\partial x}\right)\left(\frac{D}{2}\right) \quad (2.16)$$

Example 2.1 A laminar flow of oil occurs between two parallel plates that are kept 0.15 m apart at a maximum velocity of 3 m/s. Estimate the flow rate per metre width, the shear stress on plate, the pressure difference between two point 15 metres apart, the velocity gradient on the plate, and the velocity at 0.01 meters from the plates.

Assume μ as $2.5\ \text{N.s/m}^2$.

Solution

Given data:

Maximum velocity, v_{max} =3 m/s

Distance between two plates $D = 0.15$ m

Viscosity of oil $\mu = 2.5\ \text{N.s/m}^2$

Length, $L = 15$ m

Mean velocity of flow V is given by

$$V = \frac{2}{3}(3) = 2\ \text{m/s}$$

The discharge q per metre width the plate is given by

$$q = VD$$

$$q = (2 \times 0.15) = 0.3\text{m}^3/\text{s per m}$$

$$V = \frac{D^2}{12\mu}\left(-\frac{\partial p}{\partial x}\right)$$

$$\left(-\frac{\partial p}{\partial x}\right) = \frac{12\mu V}{D^2}$$

$$\left(-\frac{\partial p}{\partial x}\right) = \frac{12 \times 2.5 \times 2}{(0.15)^2} = 2666.67\text{N/m}^2 \text{ per m } \textbf{Ans.}$$

The maximum shear stress at the plate is given by

$$\tau_{\max} = \left(-\frac{\partial p}{\partial x}\right)\frac{D}{2}$$

$$\tau_{\max} = \frac{2666.7 \times 0.15}{2} = 200\ \text{N/m}^2\ .\ \textbf{Ans.}$$

The pressure difference between the two points is given by

$$(p_1 - p_2) = \frac{12\mu VL}{D^2}$$

$$(p_1 - p_2) = \frac{12 \times 2.5 \times 2 \times 15}{(0.15)^2}$$

$$(p_1 - p_2) = 40000\ \text{N/m}^2 = 40\ \text{kN/m}^2\ \textbf{Ans.}$$

The shear stress will be maximum at the plate and is given by

$$\tau_{\max} = \mu\left(\frac{\partial v}{\partial y}\right)_{y=0}$$

EXAMPLE 2.1

EXAMPLE 2.1

As such the velocity gradient at the plates is given by

$$\left(\frac{\partial v}{\partial y}\right)_{y=0} = \frac{\tau_{\max}}{\mu}$$

$$\left(\frac{\partial v}{\partial y}\right)_{y=0} = \frac{200}{2.5} = 80\ \mathrm{s}^{-1}. \textbf{ Ans.}$$

The velocity v at a distance of 0.02 m from the plate is given by

$$v = \frac{1}{2\mu}\left(-\frac{\partial p}{\partial x}\right)(Dy - y^2)$$

$$v = \frac{1}{2 \times 2.5} \times (2666.7) \times [(0.15 \times 0.01) - (0.01)^2]$$

$$v = 0.746\ \mathrm{m/s}. \textbf{ Ans.}$$

EXAMPLE 2.2

Example 2.2 Oil flows between two parallel plates separated by 150 mm at a maximum speed of 2.5 meters per second. Calculate:

(i) The discharge per metre width

(ii) The shear stress at the plates

Solution.

Given data:

Distance between plates, $D = 150$ mm

Maximum velocity, $v_{\max} = 2.5\ \mathrm{m/s}$

Viscosity of oil, $\mu = 24.5$ poise $= 2.45\ \mathrm{Ns/m^2}$

In this case the average velocity of flow,

$$V = \frac{2}{3} v_{\max}$$

$$V = \frac{2}{3} \times 2.5$$

$$V = 1.66\ \mathrm{m/s}$$

The discharge per metre width is given by

$$q = V \times D$$

$$q = 1.66 \times 0.15 = 0.249\ \mathrm{m^3/s} \text{ per m. } \textbf{Ans.}$$

The shear stress at the plates τ_0 is given by

$$q = \frac{D^3}{12\mu}\left(-\frac{\partial p}{\partial x}\right)$$

Substituting the values,

$$0.249 = \frac{0.15^3}{12 \times 2.45}\left(-\frac{\partial p}{\partial x}\right)$$

$$\left(-\frac{\partial p}{\partial x}\right) = \frac{0.249 \times 12 \times 2.45}{0.15^3}$$

$$\left(-\frac{\partial p}{\partial x}\right) = 2169.06 \text{ N/m}^2\text{/m}$$

The shear stress across any section is given by:

$$\tau = \frac{1}{2}\left(-\frac{\partial p}{\partial x}\right)(\text{D} - 2y)$$

The shear stress at the plate will be maximum and is obtained by putting $y = 0$ in the above equation. Thus,

$$\tau_{\text{max}} = \frac{1}{2}\left(-\frac{\partial p}{\partial x}\right) D$$

$$\tau_{\text{max}} = \frac{1}{2} \times 2169.06 \times 0.15$$

$$\tau_{\text{max}} = 162.67 \text{ N/m}^2.$$ **Ans.**

EXAMPLE 2.2

2.3 COUETTE FLOW

The concept of Couette flow deals with the fluid dynamics and it can be defined as the flow of viscous fluid in the space between two surfaces having one surface is fixed and another in motion. The relative motion of the two surfaces imposes in a shear stress on the fluid and surfaces. One of the classical examples of Couette flow is flow in lightly loaded in journal bearings. In order to derive the expression for a Couette flow, let us consider the velocity distribution is derived in Poisuielle flow as expressed

$$v = \frac{1}{\mu}\left(\frac{\partial p}{\partial x}\right)\frac{y^2}{2} + C_1 y + C_2 \tag{2.17}$$

Now, considering the above equation and the condition that one plate is moving with a velocity of v as shown in Fig. 2.2.

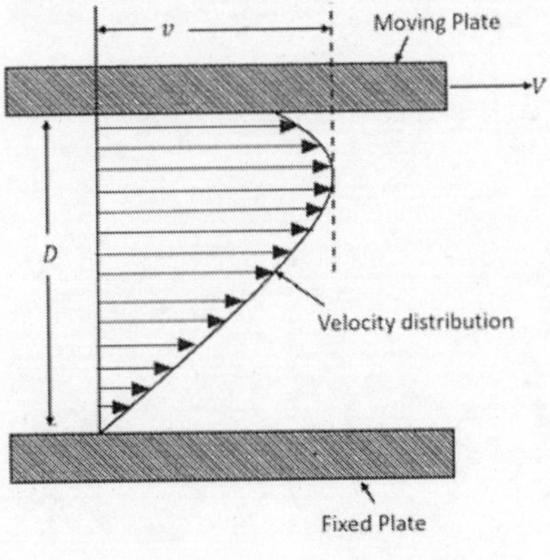

Fig. 2.2: Concept of Couette flow

Further, in this case following boundary conditions are considered for Couette flow

$$v = 0 \; at \; y = 0$$

Putting this condition in Eq. 2.17

$$C_2 = 0$$

and $$v = V \; at \; y = D$$

Putting this condition in Eq. 2.17

$$C_1 = \frac{V}{D} - \frac{1}{2\mu}\left(\frac{\partial p}{\partial x}\right) D$$

Therefore, putting the values of C_1 and C_2 in Eq. 2.17, the velocity distribution for Couette flow is obtained as

$$v = \frac{V}{D}y - \frac{1}{2\mu}\left(\frac{\partial p}{\partial x}\right)(Dy - y^2) \tag{2.18}$$

The above Equation shows that velocity distribution is the function of both V and $\left(\frac{\partial p}{\partial x}\right)$. Further, it is observed that the pressure gradient $\left(\frac{\partial p}{\partial x}\right)$ may have both positive or negative values.

For $\left(\frac{\partial p}{\partial x}\right) = 0$, the velocity distribution, v becomes linear which can be expressed as

$$v = V\frac{y}{B} \tag{2.19}$$

This Eq. 2.19 represents a simple shear flow.

Now, in order to derive the expression for the distribution of shear stress across a given section for Couette flow by considering the Newton's law of viscosity as expressed by Eq. 2.20.

$$\tau = \mu\frac{\partial v}{\partial y} \tag{2.20}$$

Substituting Eq 2.18 in Eq.2.20, we get

$$\tau = \mu\frac{\partial}{\partial y}\left[\frac{V}{D}y - \frac{1}{2\mu}\left(\frac{\partial p}{\partial x}\right)(Dy - y^2)\right] \tag{2.21}$$

or,

$$= \mu\frac{V}{D} + \left(-\frac{\partial p}{\partial x}\right)\left(\frac{D}{2} - y\right) \tag{2.22}$$

At $y = 0$; $\tau = \tau_1$

$$\tau_1 = \left[\mu\frac{V}{D} + \left(-\frac{\partial p}{\partial x}\right)\frac{D}{2}\right] \tag{2.23}$$

At middle i.e., $y = \frac{D}{2}$; $\tau = \tau_2$

$$\tau_2 = \mu\frac{V}{D} \tag{2.24}$$

At $y = D$; $\tau = \tau_3$

$$\tau_3 = \left[\mu\frac{V}{D} - \left(-\frac{\partial p}{\partial x}\right)\frac{D}{2}\right] \tag{2.25}$$

The shear stresses in Couette flow i.e., τ_1, τ_2, τ_3 at different locations are shown in Figure 2.3.

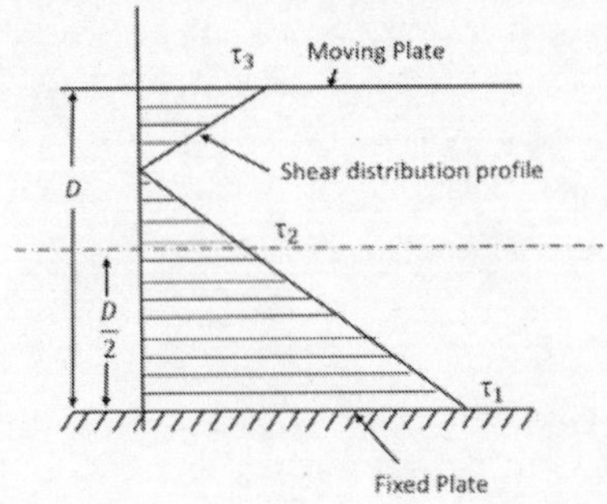

Fig. 2.3: Concept of shear stress distribution

Example 2.3 For a fluid flow, pressure distribution with respect to distance for one parallel plate having 1 mm distance and moving over the other plate is given as $\frac{\partial p}{\partial x} = -150 \times 10^6\ \text{N/m}^3$. Find out quantity of fluid over one unit of plate length and its direction if the relative velocity of the moving plate is 1.5 m/sec and viscosity of fluid, $\mu = 0.45$ poise.

Solution.

Given data:

Mean velocity, V= 1.5 m/s;.

Pressure gradient, $\frac{dp}{dx} = -150 \times 10^6\ \text{N/m}^3$

Viscosity of oil, $\mu = 0.45$ poise $= 0.045\ \text{Ns/m}^2$

Distance between plates, D= 1 mm = 0.001 m

We know that,

$$q = V \cdot \frac{D}{2} - \frac{D^3}{12\mu} \cdot \frac{\partial p}{\partial x}$$

Substituting the values

$$q = -1.5 \times \frac{0.001}{2} - \frac{0.001^3}{12 \times 0.045} \times (-150 \times 10^6)$$

$$\text{q} = 0.2769\ \text{m}^3/\text{s}$$

Hence, amount of flow per metre width = $0.268\ \text{m}^3/\text{s}$. **Ans.**

Positive direction (i.e. in the direction opposite to that of the moving plate. **Ans.**

EXAMPLE 2.3

Example 2.4 Two horizontal plates are separated by 15 mm and filled with liquid of viscosity 1 poise. There is a pressure difference of 70 kN/m^2 between two sections 55 m apart when the upper plate moves at 2.5 m/s in relation to the lower plate, which is stationary. determine:

(i) The velocity distribution,

(iii) The shear stress on the upper plate.

Solution.

Given data:

Viscosity of oil, $\mu = 1$ poise $= 0.1\ Ns/m^2$

Distance between plates, $D = 15$ mm $= 0.015$ m

Mean velocity, V= 2.5 m/s

Pressure difference $(\partial p) = 70 kN/m^2$

$\partial x = 55$ m

$$\left(-\frac{\partial p}{\partial x}\right) = \frac{70 \times 10^3}{55}$$

$$\left(-\frac{\partial p}{\partial x}\right) = 1272.73\ N/m^2/m$$

The system corresponds to Couette flow for which the velocity distribution is given as:

$$v = \frac{V}{D}y + \frac{1}{2\mu}\left(-\frac{\partial p}{\partial x}\right)(Dy - y^2)$$

$$v = \frac{2.5}{0.015}y + \frac{1}{2 \times 0.1} \times 10^3(0.015 \times y - y^2)$$

$$v = y(166.67 + 75 - 5000y)$$

$$v = y(241.67 - 5000y)$$

Hence, the velocity distribution is

$$v = y(241.67 - 5000y)$$

The shear stress on the upper plate, τ is give n by

$$\tau = \mu\left(\frac{\partial v}{\partial y}\right) = \mu\frac{\partial}{\partial y}(241.67y - 5000y^2)$$

$$\tau = 0.1(241.67 - 10000y)$$

For the top plate, $y = 0.015$ m

$$\tau_{y=0.015} = 9.167\ N/m^2.\ \textbf{Ans.}$$

EXAMPLE 2.4

2.4 LAMINAR FLOW THROUGH IN CIRCULAR CONDUITES

The laminar flow through circular conduits or pipes can be analyzed with the discussion of velocity distribution, flow rate (discharge) and the pressure drop. Consider a horizontal pipe having inside diameter aș D, as shown in Figure 2.4. A cylindrical fluid element having a radius R and length dx is considered in steady and uniform flow condition movement. The forces acting on the element on both faces with shear resistance on the periphery in the direction opposite to that of the flow.

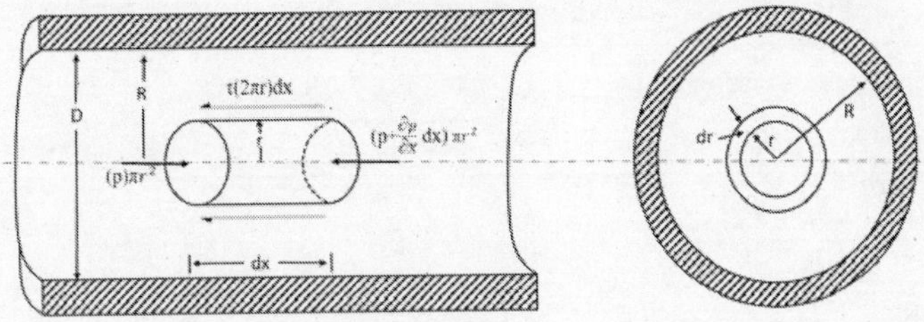

Fig. 2.4: Concept of laminar flow through in circular pipe

Let, p is the pressure intensity on the left face and τ is shear stress on the periphery of the element.

Therefore,

$\left(p + \frac{\partial p}{\partial x} dx\right)$ is pressure intensity on the right face,

$(p)\pi r^2$ is total pressure forces on the left face,

$\left(p + \frac{\partial p}{\partial x} dx\right)\pi r^2$ is total pressure forces on the right face,

$\tau(2\pi r)dx$ is total shear force acting on the periphery of the cylindrical element.

As sum of the forces acting on fluid element in the x-direction is zero.

SCAN ME

For better understanding laminar flow in circular pipe

$$p(\pi r^2) - \left(p + \frac{\partial p}{\partial x} dx\right)\pi r^2 - \tau(2\pi r)dx = 0 \tag{2.26}$$

$$\left(-\frac{\partial p}{\partial x} dx\right)\pi r^2 - \tau(2\pi r)dx = 0 \tag{2.27}$$

Shear stress,

$$\tau = -\frac{\partial p}{\partial x}\frac{r}{2} \tag{2.28}$$

It can be seen from figure that the shear stress will be zero and maximum corresponding to r=0 and $r = R$, respectively. The maximum value of shear stress can be expressed as

$$\tau_{max} = \left(-\frac{\partial p}{\partial x}\right)\frac{R}{2} \tag{2.29}$$

The negative sign indicates the decrease in pressure of fluid in the flow direction. Eq. 2.29 represents the maximum shear stress for laminar flow through pipes.

For the uniform pipe, pressure gradient will be constant and can be expressed as

$$\left(-\frac{\partial p}{\partial x}\right) = \left(\frac{p_1 - p_2}{L}\right) = \frac{\rho g h_f}{L} \tag{2.30}$$

Substituting Eq. 2.30 in Eq. 2.28, the shear stress can be expressed as

$$\tau = \frac{\rho g h_f}{2L} r \tag{2.31}$$

Where, h_f is the head loss due to friction through pipes.

From the Newton's law of viscosity, $\left(\tau = -\mu \frac{\partial v}{\partial r}\right)$ and Eq. 2.28, we get

$$-\mu \frac{\partial v}{\partial r} = -\frac{\partial p}{\partial x}\frac{r}{2} \tag{2.32}$$

$$\frac{\partial v}{\partial r} = \frac{1}{\mu}\frac{\partial p}{\partial x}\frac{r}{2} \tag{2.33}$$

By integrating Eq. 2.33, the expression for the velocity distribution for laminar flow in a circular pipe can be written as

$$v = \frac{1}{4\mu}\left(\frac{\partial p}{\partial x}\right) r^2 + C \tag{2.34}$$

By taking the boundary condition as velocity $v = 0$ for $r = R$, the constant C is determined as

$$C = -\frac{1}{4\mu}\left(\frac{\partial p}{\partial x}\right) R^2 \tag{2.35}$$

Finally, the velocity distribution of laminar flow through pipes is expressed as

$$v = \frac{1}{4\mu}\left(-\frac{\partial p}{\partial x}\right)(R^2 - r^2) \tag{2.36}$$

By applying the boundary condition, velocity $v = v_{max}$ for $r = 0$,

$$v_{max} = \frac{1}{4\mu}\left(-\frac{\partial p}{\partial x}\right) R^2 \tag{2.37}$$

Dividing Eq. 2.36 by Eq. 2.37, the expression for the velocity distribution can be written in terms of the maximum velocity as

$$v = v_{\max}\left[1 - \left(\frac{r}{R}\right)^2\right] \tag{2.38}$$

Fig. 2.5 shows the profile velocity distribution based on Eq. 2.36 and shear stress distribution based on Eq. 2.28.

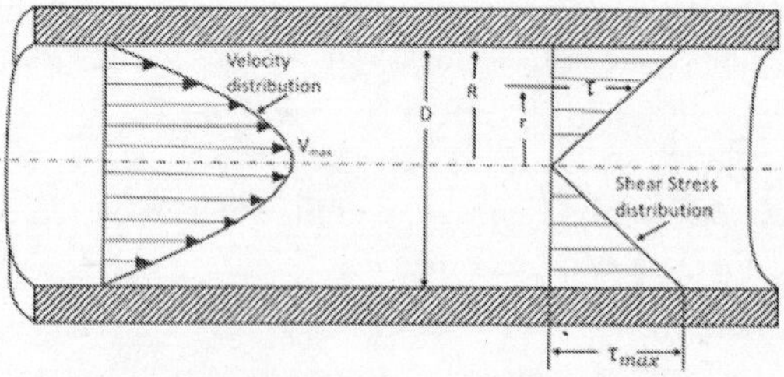

Fig. 2.5: Velocity and shear stress distribution profiles for laminar flow through circular pipe

Based on the velocity distribution concept for laminar flow in circular pipe, the discharge (flow rate) can be discussed as follows

Let us consider an elementary ring having thickness dr and discharge passing through which can be expressed as

$$dQ = vdA = v(2\pi r)dr \tag{2.39}$$

$$dQ = \frac{1}{4\mu}\left(-\frac{\partial p}{\partial x}\right)(R^2 - r^2)(2\pi r)dr \tag{2.40}$$

The total discharge is obtained by integrating Eq. 2.40 at r = R and can be expressed as

$$Q = \frac{\pi}{2\mu}\left(-\frac{\partial p}{\partial x}\right)\frac{R^4}{4} \tag{2.41}$$

$$Q = \frac{\pi}{8\mu}\left(-\frac{\partial p}{\partial x}\right)R^4 \tag{2.42}$$

$$Q = \frac{\pi}{128\mu}\left(-\frac{\partial p}{\partial x}\right)D^4 \tag{2.43}$$

Where, D is the diameter of pipe.

From continuity equation, mean velocity of flow can be determined as

$$V = \frac{Q}{A} \tag{2.44}$$

Where, $A = \frac{\pi}{4}D^2$

Putting Q and A values in Eq. 2.44, Thus

$$V = \frac{1}{32\mu}\left(-\frac{\partial p}{\partial x}\right)D^2 \tag{2.45}$$

From Eq. 2.45 and 2.37,

$$V = \frac{1}{2}v_{\max} \tag{2.46}$$

If local velocity is equal to the mean velocity of flow V,

Then

$$V = v_{\max}\left[1 - \left(\frac{r}{R}\right)^2\right] = \frac{1}{2}v_{\max} \tag{2.47}$$

$$r = 0.707R \tag{2.48}$$

Therefore, the mean velocity of flow V occurs at a radial distance of $(0.707R)$ from the centre of the pipe.

Further, using Eq. 2.30 the pressure drop can be expressed as

$$\left(-\frac{\partial p}{\partial x}\right) = \frac{32\mu V}{D^2} \tag{2.49}$$

Consider the pressure drop occurs along a pipe length of L,

$$\frac{p_1 - p_2}{L} = \frac{32\mu V}{D^2} \tag{2.50}$$

or,

$$p_1 - p_2 = \frac{32\mu VL}{D^2} \tag{2.51}$$

$$p_1 - p_2 = \frac{128\mu QL}{\pi D^4} \tag{2.52}$$

The above Eq. 2.51 is known as Hagen-Poiseuille equation for laminar flow in the circular pipe.

Further, friction losses (head losses) in circular pipe for a laminar flow over pipe length, L can be expressed as

$$h_f = \frac{p_1 - p_2}{\rho g} = \frac{32\mu VL}{\rho g D^2} \tag{2.53}$$

Putting Reynolds number ($\text{Re} = \rho VD/\mu$) in Eq. 2.53

$$h_f = \frac{64}{\text{Re}}\left(\frac{LV^2}{2gD}\right) \tag{2.54}$$

Or

$$h_f = \left(\frac{128\mu QL}{\rho g\pi D^4}\right) \tag{2.55}$$

The head loss due to frictional resistance in a long straight pipe of length L and diameter D may also be expressed by Darcy-Weisbach equation as

$$h_f = \frac{fLV^2}{2gD} \tag{2.56}$$

where f is friction factor and V is mean velocity of flow.

By comparing Eq. 2.56 and Eq. 2.54, the friction factor for a laminar flow in circular piper can be given as

$$f = \frac{64\mu}{\rho VD} = \frac{64}{\text{Re}} \tag{2.57}$$

Further, the expression for maximum shear stress can be determined as follows,

$$\tau_{max} = \left(-\frac{\partial p}{\partial x}\right)\frac{R}{2} \tag{2.58}$$

$$= \left(\frac{p_1 - p_2}{L}\right)\frac{D}{4} \tag{2.59}$$

Where, $(p_1 - p_2) = \rho g h_f$

Thus,

$$\tau_{max} = \frac{(\rho g h_f)D}{4L} = \frac{\rho f V^2}{8}$$

$$\frac{\tau_{max}}{\rho} = \frac{fV^2}{8} \tag{2.60}$$

Or

$$\sqrt{\frac{\tau_{max}}{\rho}} = V\sqrt{\frac{f}{8}} = V_* \tag{2.61}$$

Where, V_* expressed as $\left[\sqrt{(\tau_{\max}/\rho)}\right]$ is the dimensions of velocity and known as shear velocity, V_*.

The power (P) required to overcome the flow resistance to maintain steady uniform laminar flow through pipes can be determined as

$$P = \left(-\frac{\partial p}{\partial x}\right) ALV = Q(p_1 - p_2) \tag{2.62}$$

Where,

$-\partial p/\partial x$ is the pressure gradient in the direction of flow,

A is the area of pipe,

L is the length of pipe,

V is the man velocity of flow.

2.5 LAMINAR FLOW THROUGH ANNULUS

The Fig. 2.6 shows a steady laminar flow between two concentric circular tubes in an annular space. Consider fluid elements are subjected to normal pressure forces at the ends and shear forces at the inner and outer curves of their curved surfaces in the direction of flow.

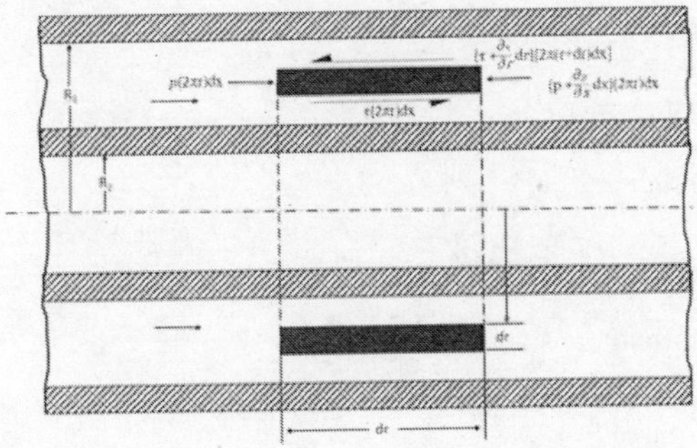

Fig. 2.6: Concept of laminar flow through an annulus

Let,

R_1 is the inner surface radius of outer tube

R_2 is the outer surface radius of inner tube

dx is the length of small concentric cylindrical sleeve of the fluid element

dr is the thickness of small concentric cylindrical sleeve of the fluid element

r is the radial distance

As sum of the forces acting on fluid element in the direction of flow is zero.

$$\left[p(2\pi r)dr - \left(p + \frac{\partial p}{\partial x}dx\right)2\pi r dr\right] + \left[\tau(2\pi r)dx - \left(\tau + \frac{\partial \tau}{\partial r}dr\right) \times 2\pi(r + dr)dx\right] = 0 \tag{2.63}$$

Or

$$\left[\left(-\frac{\partial p}{\partial x}dx\right)2\pi r dr - \left(\frac{\partial \tau}{\partial r}dr\right)2\pi r dx\right] - \left[\tau(2\pi dx)dr + \left(\frac{\partial \tau}{\partial r}dr\right)(2\pi dx)dr\right] = 0 \tag{2.64}$$

Solving and neglecting higher order term in Eq. 2.64, we get

$$\left(\frac{\partial p}{\partial x} + \frac{\partial \tau}{\partial r} + \frac{\tau}{r}\right) = 0 \tag{2.65}$$

The p is dependent on x and τ is dependent on r in Eq. 2.65

$$\frac{\partial p}{\partial x} + \frac{1}{r}\frac{\partial}{\partial r}(\tau r) = 0.$$

$$r\frac{\partial p}{\partial x} + \frac{\partial}{\partial r}(\tau r) = 0. \tag{2.66}$$

Integrating Eq. 2.66 w. r. to r

$$\left(\frac{\partial p}{\partial x}\right)\frac{r^2}{2} + \tau r = C_1 \tag{2.67}$$

Substituting $\tau = -\mu(\partial v/\partial r)$ in Eq. 2.67.

$$\left(\frac{\partial p}{\partial x}\right)\frac{r^2}{2} - \mu r\frac{\partial v}{\partial r} = C_1 \tag{2.68}$$

Dividing Eq. 2.68 by r and integrating w. r. to r,

$$\left(\frac{\partial p}{\partial x}\right)\frac{r^2}{4} - \mu v = C_1 \log_e r + C_2 \tag{2.69}$$

Where, C_1 and C_2 are the second constant of integration.

Applying the boundary conditions in Eq. 2.69 at $r = R_1$ and $r = R_2, v = 0$. The velocity distribution in an annulus can be expressed by:

$$v = -\frac{1}{4\mu}\left(\frac{\partial p}{\partial x}\right) \times \left[R_1^2 - r^2 - \frac{(R_1^2 - R_2^2)}{\log_e(R_1/R_2)}\log(R_1/r)\right] \tag{2.70}$$

The location for maximum velocity can be obtained by differentiate Eq. 2.70, w. r. to r and equate it to zero.

$$\frac{\partial v}{\partial r} = 0 = -\frac{1}{4\mu}\left(\frac{\partial p}{\partial x}\right)\left[-2r + \frac{1}{r}\frac{(R_1^2 - R_2^2)}{\log_e(R_1/R_2)}\right] \tag{2.71}$$

Rearranging Eq. 2.71, we get

$$r = \left[\frac{R_1^2 - R_2^2}{2\log_e(R_1/R_2)}\right]^{1/2} \tag{2.72}$$

The maximum velocity can be obtained by substituting the value of r in Eq. 2.70.

The total discharge passing through the cross of the annulus can be expressed as

$$Q = \int_{R_2}^{R_1} 2\pi r v dr \tag{2.73}$$

Substituting the value of v in Eq. 2.73 and integrating it w. r. to r

$$Q = \left(-\frac{\pi}{8\mu}\right)\left(\frac{\partial p}{\partial x}\right)\left[R_1^4 - R_2^4 - \frac{(R_1^2 - R_2^2)^2}{\log_e(R_1/R_2)}\right] \tag{2.74}$$

The mean velocity of flow through the annulus is obtained by dividing Eq. 2.74 by $\pi(R_1^2 - R_2^2)$

$$V = \left(-\frac{1}{8\mu}\right)\left(\frac{\partial p}{\partial x}\right)\left[\left(R_1^2 + R_2^2 - \frac{(R_1^2 - R_2^2)}{\log_e(R_1/R_2)}\right)\right] \tag{2.75}$$

The mean velocity for laminar flow through circular pipe is obtained by putting the value of $R_2 = 0$ in Eq. 2.76.

For an inclined annulus, the velocity distribution and discharge expressions are obtained by replacing with $(\partial p/\partial x)$ by $[\partial(p + wz)/\partial x)$.

Example 2.5 A pipe of diameter of 35 mm is inclined upward at an angle of 60^0 from the horizontal plane and covering water with mean velocity of 2 m/s. Two sections down side the pipe is marked keeping distance of 25 m between these marks. If the pressure corresponding to these marks are 600 kPa and 300 kPa; find out the shear stress at the wall and at a radius of 8 mm.

Solution

Given data:

$$D = 0.035 \text{ mm}$$

$$R = \frac{D}{2} = 0.0175\, m$$

$$p_1 = 600 \text{ kpa}$$

$$p_2 = 300 \text{ kpa}$$

$$r = 8 \text{ mm}$$

$$V = 2 \text{ m/s}$$

Assuming datum to be passing through the lower point, we have

$$(h_1 - h_2) = \left(\frac{p_1}{w} + 0\right) - \left(\frac{p_2}{w} + Z_2\right)$$

$$(h_1 - h_2) = \left(\frac{600 \times 10^3}{9810}\right) - \left(\frac{300 \times 10^3}{9810} + \frac{25}{\sqrt{2}}\right)$$

$$(h_1 - h_2) = (61.162 - 48.258)$$

$$(h_1 - h_2) = 12.904 \text{ m}$$

EXAMPLE 2.5

EXAMPLE 2.5

Average shear stress at the wall of the pipe is given by

$$\tau_{\max} = \rho g\left(-\frac{\partial h}{\partial x}\right)\frac{R}{2}$$

$$\tau_{\max} = \frac{9810 \times 12.904}{25} \times \frac{0.0175}{2}$$

$\tau_{\max} = 44.305\ \text{N/m}^2$. **Ans.**

Average shear stress at a radius of 8 mm is given by

$$\tau = \left(-\frac{\partial h}{\partial x}\right)\frac{r}{2}$$

$$\tau = \frac{9810 \times 12.904}{25} \times \frac{0.008}{2}$$

$\tau = 20.254\ \text{N/m}^2$. **Ans.**

EXAMPLE 2.6

Example 2.6 Determine the pressure drop in a pipe through which an oil (viscosity = 0.09 Ns/m², relative density 0.88) is flowing for the following data;

(i) Dia. of pipe : 60 mm
(ii) Length of pipe : 400 m
(iii) Flow rate : 4 lps

Solution

Given data:

$$\mu = 0.09\ \text{Ns/m}^2$$

$$\rho_0 = 0.88 \times \rho_w$$

$$\rho_0 = 0.88 \times 1000 = 880\ \text{kg/m}^3$$

$$D = 60\ \text{mm} = 0.06\ \text{m}$$

$$L = 400\ \text{m}$$

$$Q = 0.004\ \text{m}^3/\text{s}$$

$$V = \frac{Q}{\text{Area}}$$

$$V = \frac{0.004}{\frac{\pi}{4}0.06^2}$$

$$V = 1.414\ \text{m/s}$$

The Reynolds number (R_e) is given by,

$$R_e = \frac{\rho V D}{\mu}$$

$$R_e = 880 \times \frac{1.414 \times 0.06}{0.09}$$

$$R_e = 829.546$$

As Reynolds number is less than 2000, the flow is viscous or laminar

So the pressure drop is given by

$$p_1 - p_2 == \frac{32\mu VL}{D^2}$$

$$p_1 - p_2 \quad = \frac{32 \times 0.09 \times 1.414 \times 400}{(.06)^2}$$

$$p_1 - p_2 = 452480 \text{ N/m}^2$$

$p_1 - p_2 = 45.248 \text{ N/cm}^2$. **Ans.**

EXAMPLE 2.6

Example 2.7 Determine the pressure gradient, the average velocity for a fluid flowing through a pipe having diameter of 150 mm. Take the oil properties and other conditions as follows;

(i)	Viscosity	: 0.8 Ns/m^2
(ii)	Specific gravity	: 1.4
(iii)	Maximum shear stress at wall	: 200 N/m^2

Solution

Given data:

$\mu = 0.8$ Ns/m^2

Specific gravity =1.4

$\rho = 1.4 \times 1000 = 1400$ kg/m^3

$D = 150$ mm = 0.15 m

$$R = \frac{D}{2} = 0.075 \text{ m}$$

$\tau_{max} = 200$ N/ m^2

The maximum shear stress (τ_{max}) is given by

$$\tau_{max} = -\frac{\partial p}{\partial x}\frac{R}{2}$$

$$200 = -\frac{\partial p}{\partial x} \times \frac{0.075}{2}$$

$\frac{\partial p}{\partial x} = -5333.33 \text{ N/m}^2$per m **Ans.**

(ii) Average velocity

$$V \ = \frac{1}{2} v_{max}$$

$$V = \frac{1}{2}\left[-\frac{1}{4\mu}\frac{\partial p}{\partial x}R^2\right]\left\{\because v_{max} = -\frac{1}{8\mu}\frac{\partial p}{\partial x}R^2\right\}$$

$$V = \frac{1}{8\mu} \times \left(-\frac{\partial p}{\partial x}\right) R^2$$

$$V = \frac{1}{8 \times 0.8} \times (5333.33) \times (0.075)^2$$

$V = 4.687$ m/s. **Ans.**

EXAMPLE 2.7

2.6 CONCEPT OF BOUNDARY LAYER

As discussed in previous unit, that all the fluids even having very low viscosity are considered real fluids. When real fluids flowing in contact with a sold surface, the fluid particles adhere to the boundary of the solid surface due to viscosity of fluids and the fluid velocity at the boundary surface will be zero. Under this situation, a no-slip condition occurs as the relative motion of fluid with plate becomes zero at the boundary. Therefore, in the vicinity of the solid surface there will be variation in velocity from zero to fluid flow mean velocity in a region. It is due to fluid layer at the surface retards the adjacent layer and it will continue up to the layer of the fluid which is moving with the fluid flow mean velocity. This region of layers is known as boundary layer. In the boundary layer region, a large velocity gradient $\left(\frac{\partial v}{\partial y}\right)$ will occur and due to this velocity gradient, there will be a shear stress, $\tau = \mu\left(\frac{\partial v}{\partial y}\right)_{y=0}$ in this region of boundary layer. The force caused by this stress in the direction of motion is known as surface drag or drag. The wall in tern exerts an equal force on the fluid in opposite direction will be known as shear resistance to the flow as shown in Fig. 2.7.

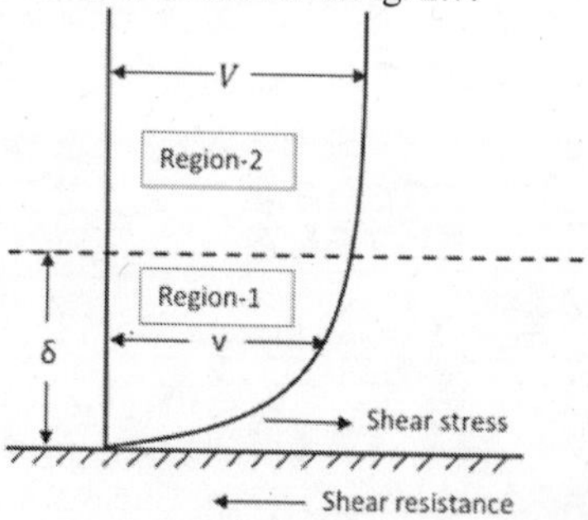

SCAN ME

For better understanding boundary layer

Fig. 2.7: Concept of boundary layer

A seen in the figure that two different regions as viscous and non-viscous exist which are separated by the boundary layer. The viscous region with in the boundary layer has larger velocity gradient due to viscous forces however non-viscous region outside the boundary layer experiences negligible viscous forces. *L*. Prandtl introduced the boundary layer concept in 1904 and since then it has been applied to several fluid flow applications.

2.7 THICKNESS OF BOUNDARY LAYER

As discussed in section 2.6, due to boundary layer two different regions exist i.e., viscous and non-viscous. Within viscous region the velocity increases from zero at the boundary surface to the velocity of the main stream asymptotically. The thickness of this region is known as boundary layer thickness and represented by $'\delta'$ as shown in Fig. 2.10. As transition of the velocity takes place asymptotically, it is difficult to define the thickness of the boundary layer. It is therefore, an arbitrarily boundary layer thickness can be defined as boundary layer thickness and at $y = \delta$, $v = 0.99V$ it is also known as nominal thickness of boundary layer.

In order to achieve the better accuracy, the boundary layer thickness can be mathematically defined as displacement thickness δ^* , the momentum thickness θ and the energy thickness δ_E.

2.7.1 Displacement thickness

In boundary layers, displacement thickness, δ^* refers to the distance in which the surface should be displaced perpendicular to the fluid flow (y-direction) in order to compensate for the loss of flow rate resulting from boundary layer formation. Hence, it can be expressed as

$$\frac{\partial p}{\partial x} = \mu \frac{d^2 v}{dy^2} \tag{2.76}$$

$$V\delta^* = \int_0^\infty (V - v)dy$$

Or

$$\delta^* = \int_0^\infty \left(1 - \frac{v}{V}\right) dy \tag{2.77}$$

2.7.2 Momentum thickness

In boundary layers, momentum thickness, θ refers to the distance in which the surface should be displaced perpendicular to the fluid flow (y-direction) in order to compensate for the loss of momentum of fluid flow resulting from boundary layer formation.

Thus,

$$\rho V^2 \theta = \rho \int_0^\infty (V - v)v dy \tag{2.78}$$

Or

$$\theta = \int_0^\infty \frac{v}{V}\left(1 - \frac{v^2}{V^2}\right) dy \tag{2.79}$$

2.7.3 Energy thickness

In boundary layers, the energy thickness, δ_E refers to the distance in which the surface should be displaced perpendicular to the fluid flow (y-direction) in order to compensate for the loss of kinetic energy resulting from boundary layer formation.

Thus,

$$\frac{1}{2}\rho V^3 \delta_E = \frac{1}{2}\rho \int_0^\infty (V^2 - v^2)v dy \tag{2.80}$$

Or

$$\delta_E = \int_0^\infty \frac{v}{V}\left(1 - \frac{v^2}{V^2}\right) dy \tag{2.81}$$

Example 2.8 Determine the displacement, momentum, and energy thickness for a boundary layer of $\frac{u}{U} = \frac{y}{\delta}$ where u represents the velocity of the plate at a distance of y from it and $u = U$ at $y = \delta, \delta$ is the boundary layer thickness.

Solution

(i) The displacement thickness, δ^*

$$\delta^* = \int_0^\delta \left(1 - \frac{u}{U}\right) dy$$

$$\delta^* = \int_0^\delta \left(1 - \frac{y}{\delta}\right) dy \ \left(\because \frac{u}{U} = \frac{y}{\delta}\right)$$

$$\delta^* = \left[y - \frac{y^2}{2\delta}\right]_0^\delta$$

$$\delta^* = \left(\delta - \frac{\delta^2}{2\delta}\right) = \delta - \frac{\delta}{2} = \frac{\delta}{2}. \textbf{ Ans.}$$

(ii) The momentum thickness, θ

$$\theta = \int_0^\delta \frac{u}{U}\left(1 - \frac{u}{U}\right) dy$$

$$\theta = \int_0^\delta \frac{y}{\delta}\left(1 - \frac{y}{\delta}\right) dy = \int_0^\delta \left(\frac{y}{\delta} - \frac{y^2}{\delta^2}\right) dy$$

$$\theta = \left[\frac{y^2}{2\delta} - \frac{y^3}{3\delta^2}\right]_0^\delta = \frac{\delta^2}{2\delta} - \frac{\delta^3}{3\delta^2} = \frac{\delta}{2} - \frac{\delta}{3} = \frac{\delta}{6}. \textbf{ Ans.}$$

(iii) The energy thickness, δ_e

$$\delta_e = \int_0^\delta \frac{u}{U}\left(1 - \frac{u^2}{U^2}\right) dy$$

$$\delta_e = \int_0^\delta \frac{y}{\delta}\left(1 - \frac{y^2}{\delta^2}\right) dy = \int_0^\delta \left(\frac{y}{\delta} - \frac{y^3}{\delta^3}\right) dy$$

$$\delta_e = \left[\frac{y^2}{2\delta} - \frac{y^4}{4\delta^3}\right]_0^\delta = \frac{\delta^2}{2\delta} - \frac{\delta^4}{4\delta^3} = \frac{\delta}{2} - \frac{\delta}{4} = \frac{\delta}{4}$$

$$\delta_e = \frac{\delta}{4}. \textbf{ Ans.}$$

EXAMPLE 2.8

Example 2.9 A velocity distribution in the boundary layer can be expressed as follows:

$$\frac{v}{V} = \frac{3}{2}\eta - \frac{1}{2}\eta^2$$

If $\eta = (y/\delta)$, determine (δ^*/δ) and (θ/δ).

Solution

$$\delta^* = \int_0^\infty \left(1 - \frac{v}{V}\right) dy$$

$$\delta^* = \int_0^\delta \left(1 - \frac{v}{V}\right) dy + \int_\delta^\infty \left(1 - \frac{v}{V}\right) dy$$

But outside the boundary layer $(v/V) = 1$, and hence

$$\delta^* = \int_0^\delta \left(1 - \frac{v}{V}\right) dy$$

$$\frac{v}{V} = \frac{3}{2}\eta - \frac{1}{2}\eta^2 \text{ and } dy = \delta d\eta$$

Thus,

$$\delta^* = \delta \int_0^1 \left(1 - \frac{3}{2}\eta + \frac{1}{2}\eta^2\right) d\eta$$

$$\delta^* = \frac{5}{12}\delta$$

$$\frac{\delta^*}{\delta} = \frac{5}{12}. \textbf{ Ans.}$$

Similarly,

$$\theta = \int_0^\delta \frac{v}{V}\left(1 - \frac{v}{V}\right) dy$$

$$\theta = \delta \int_0^1 \left(\frac{3}{2}\eta - \frac{1}{2}\eta^2\right)\left(1 - \frac{3}{2}\eta + \frac{1}{2}\eta^2\right) d\eta$$

$$\theta = \frac{19}{120}\delta$$

$$\frac{\theta}{\delta} = \frac{19}{120}. \textbf{ Ans.}$$

EXAMPLE 2.9

Example 2.10 Find out the maximum distance for air flowing over a flat plate at wind velocity of 100 km/h, if the laminar boundary layer exists up to Reynold number equal to 2.5 $\times 10^5$. Use the standard value of kinematic viscosity of air.

Solution

Given data

$Re_x = 2.5 \times 10^5$

EXAMPLE 2.10

EXAMPLE 2.10

Assume

$V = 100$ km/hr

$v = 1.49 \times 10^{-5}\ \mathrm{m^2/s}$

The velocity in m/s is given by

$$V = \frac{100 \times 10^3}{3600}$$

The velocity in m/s is given by

$$V = \frac{100 \times 10^3}{3600}$$

$V = 27.78\ \mathrm{m/s}$

The Reynolds number is given by

$$\mathrm{Re}_x = \frac{Vx}{v}$$

$$2.5 \times 10^5 = \frac{27.78 \times x}{1.49 \times 10^{-5}}$$

$x = 0.134$ m

EXAMPLE 2.11

Example 2.11 In a fluid flow boundary layer is formed for which velocity distribution is represented by $\frac{v}{V} = \left(\frac{y}{\delta}\right)^{1/7}$. Find out the parameter $\frac{\delta^*}{\delta}, \frac{\theta}{\delta}$ and $\frac{\delta_E}{\delta}$.

Solution

Given data

$V = 15$ m/s

$\delta_E = 20$ mm

Assume

$\rho = 1.226\ \mathrm{kg/m^3}$

The displacement thickness is given by

$$\delta^* = \int_0^\delta \left(1 - \frac{v}{V}\right) dy$$

$$\delta^* = \int_0^\delta \left[1 - \left(\frac{y}{\delta}\right)^{1/7}\right] dy$$

$$\delta^* = \frac{\delta}{8}$$

$$\frac{\delta^*}{\delta} = \frac{1}{8} = 0.125$$

The momentum thickness is given by

$$\theta = \int_0^\delta \frac{v}{V}\left(1 - \frac{v}{V}\right) dy$$

EXAMPLE 2.11

$$\theta = \int_0^\delta \left(\frac{y}{\delta}\right)^{1/2}\left[1-\left(\frac{y}{\delta}\right)^{1/7}\right]dy$$

$$\theta = \frac{7}{72}\delta$$

$$\frac{\theta}{\delta} = \frac{7}{72} = 0.097$$

The energy thickness is given by

$$\delta_E \ = \int_0^\delta \frac{v}{V}\left(1-\frac{v^2}{V^2}\right)dy$$

$$\delta_E = \int_0^\delta \left(\frac{y}{\delta}\right)^{1/7}\left[1-\left(\frac{y}{\delta}\right)^{2/7}\right]dy$$

$$\delta_E = \frac{7}{40}\delta$$

$$\frac{\delta_E}{\delta} = \frac{7}{40} = 0.175$$

2.8 DARCY-WEISBACH EQUATION

In pipe flow it becomes necessary to determine the head losses to design the piping system. The Darcy-Weisbach equation is considered the only formulae to determine the head losses accurately for fluid flow through pipes. The Darcy-Welsbach equation is derived considering the pressure losses or head losses due to friction, h_L along the given length of a pipe, L through which fluid is flowing with an average velocity, V for incompressible fluid as shown in Fig. 2.8.

Let,

A is the cross-sectional area of pipe,

p_1 is pressure intensity at section 1,

v_1 is flow velocity at section 1, and

p_2, v_2 are the corresponding values at section 2, respectively.

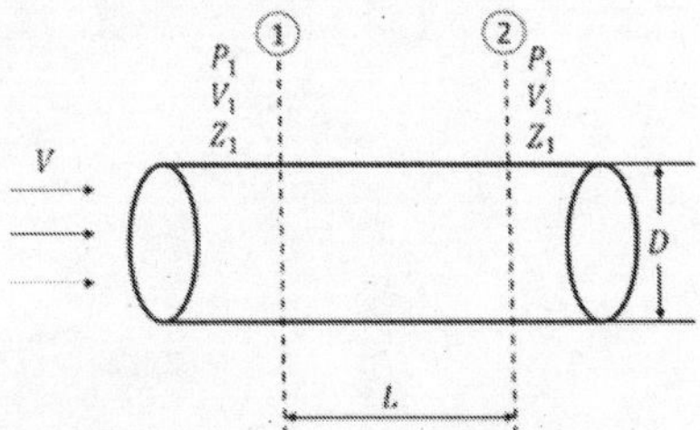

Fig. 2.8: Fluid flow through pipe

Applying Bernoulli's equation between the sections 1 and 2,

$$\frac{p_1}{\rho g}+\frac{v_1^2}{2g}+z_1 = \frac{p_2}{\rho g}+\frac{v_2^2}{2g}+z_2+h_L \tag{2.82}$$

Where, h_L is head losses due to friction

Further, due to the pipe is located in horizontally and flow is uniform,

$z_1 = z_2$ and $v_1 = v_2 = v$

Thus, Eq. 2.82 becomes

$$h_L = \frac{p_1}{\rho g} - \frac{p_2}{\rho g} \tag{2.83}$$

The above Eq. 2.83 represents difference between pressure heads at any two sections will be equal to head losses due to friction in a pipe.

Further, consider f' be the frictional resistance per unit area at unit velocity, then

$$\text{frictional resistance} = f' \times pL \times v^n \tag{2.84}$$

where, P is the wetted perimeter of the pipe and n=2.

Now, let the pressure force at section 1 and 2 are $p_1 A$ and $p_2 A$ respectively. Therefore, by balancing these forces, we get

$$p_1 A = p_2 A + f' \times pL \times v^2$$

$$(p_1 - p_2) = f' \times \frac{p}{A} \times Lv^2 \tag{2.85}$$

Now dividing the ρg on both sides in the above Eq., we get

$$\frac{p_1 - p_2}{\rho g} = \frac{f'}{\rho g} \times \frac{p}{A} \times Lv^2 \tag{2.86}$$

Comparing Eq. 2.83 and Eq. 2.86, we get

$$h_L = \frac{f'}{\rho g} \times \frac{p}{A} \times Lv^2 \tag{2.87}$$

Where, the ratio of cross-section area of pipe (A) and wetted perimeter (P) known as hydraulic mean diameter, m.

For pipes, hydraulic mean diameter (m) can be expressed as

$$m = \frac{A}{P} = \frac{\frac{\pi}{4}D^2}{\pi D}$$

$$m = \frac{D}{4}$$

Putting the value of m in Eq. 2. 87, we get

$$h_L = \frac{4f'}{\rho g} \times \frac{Lv^2}{D} \tag{2.88}$$

Head losses due to friction in a pipe can be expressed as

$$h_L = \frac{fLv^2}{2gD} \tag{2.89}$$

where f is known as *friction factor*, which is a dimensionless quantity.

The Eq. 2.89 is known as Darcy-Welsbach equation which is commonly used for calculating the head loss due to friction in pipes. To determine the head loss due to friction in pipes, the value of friction factor need to be calculated by using the various correlations developed for smooth and rough boundaries. The friction factor is the strong function of surface roughness condition and Reynolds number.

2.9 FRICTION FACTOR

As discussed above in section 2.8, the head loss in pipe flow can be determined using Darcy-Welsbach equation. These losses are due to frictional resistance between fluid and the pipe surface. Using the correct value of friction factor, it is possible to determine the amount of energy dissipated as a result of the frictional resistance in the pipe. The friction factor (f) depends upon the non-dimensional parameters such as Reynolds number ($Re = vD/\nu$) and roughness height to pipe diameter ratio (k/D). Mathematically, it can be written as

$$f = \phi\left[\left(\frac{vD}{\nu}\right), \left(\frac{k}{D}\right)\right] \tag{2.90}$$

Eq. 2.30 is applicable for both laminar and turbulent flows in pipe and in both the cases friction factor is the function of Reynolds number, Re. However, for laminar flow it depends on Reynolds number only and can be expressed by Eq. 2.57 derived under section 2.4.

$$f = \frac{64}{Re} \tag{2.91}$$

As stated previously, the friction factor for a fully developed turbulent flow depends either on Re or (k/D) or both, depending on whether boundaries are hydrodynamically rough or smooth. The friction factor relationships for smooth and rough boundaries are expressed in below sub-sections.

2.9.1 Friction Factor Variation for Smooth Pipes

A hydrodynamically smooth pipe for turbulent flow has roughness protrusions submerged in the viscous sub-layer, so the friction factor is independent on the relative roughness (k)and is only the function Reynolds number (Re). Saph and Schoder conducted an experimental analysis which is further studied by Blasius and developed the correlations for friction factor in smooth pipes and it can be expressed as

$$f = \frac{0.316}{(Re)^{1/4}} \tag{2.93}$$

In Fig. 2.9, Blasius equation is plotted using the logarithmic scale with $\log Re$ taken as X coordinate and $\log f$ taken as Y coordinate. The graph of log Re and log f shows the Blasius equation is valid for the Reynolds number varies from 4×10^3 to 10^5 . For Reynolds number greater than 10^5, the Nikuradse conducted experimental analysis to derive the expression for friction factor in smooth pipes by using the logarithmic law of velocity distribution.

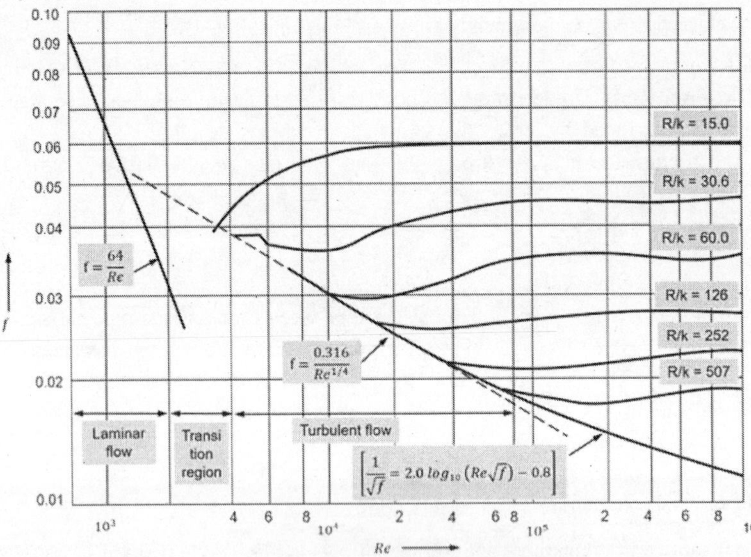

Fig. 2.9: Variation of friction factor with Reynolds number for artificially roughened pipes

The mean velocity of flow (V) for turbulent flow in smooth pipes can be expressed by

$$\frac{V}{V_*} = 5.75\log_{10}\left(\frac{V_* R}{\upsilon}\right) + 1.75 \tag{2.94}$$

Substituting $V_* = V\sqrt{\frac{f}{8}}$, in Eq. 2.94, it becomes

$$\frac{V}{V(f/8)^{1/2}} = 5.75\log_{10}\left(\frac{V(f/8)^{1/2}R}{\upsilon}\right) + 1.75 \tag{2.95}$$

By rearranging Eq. 2.95, we get

$$\frac{1}{\sqrt{f}} = 2.03\log_{10}\left(\frac{VD}{\upsilon}\sqrt{f}\right) - 0.91 \tag{2.96}$$

or

$$\frac{1}{\sqrt{f}} = 2.03\log_{10}(\text{Re}\sqrt{f}) - 0.91 \tag{2.97}$$

Based on Nikuradse's experimental values for turbulent flow in smooth pipe indicates that the experimental results follow closely the trend of the following Eq. 2.98 instead of Eq. 2.97.

$$\frac{1}{\sqrt{f}} = 2.0\log_{10}(\text{Re}\sqrt{f}) - 0.8 \tag{2.98}$$

In the case of turbulent flow in smooth pipes, Eq. 2.98 is referred to as the Karman-Prandtl resistance equation and is valid for the Reynolds number varying from 5×10^4 to 4×10^7. As f appears on both sides of Eq. 2.98, it can only be solved by trial and error. So, Nikuradse has developed the empirical relationship for the value of friction factor (f) which fits experimental data very well and can be expressed as

$$f = 0.0032 + \frac{0.221}{(\mathrm{Re})^{0.237}} \tag{2.99}$$

2.9.2 Friction Factor Variation for Rough Pipes

For turbulent flow in rough pipes, friction factor is independent on value of Reynolds number (Re) and is the strong function of relative roughness (k). The expression for friction factor in rough pipes can be obtained by considering the mean velocity of flow (v) and can be expressed as

$$\frac{V}{V_*} = 5.75\log_{10}\left(\frac{R}{k}\right) + 4.75 \tag{2.100}$$

Substituting $V_* = V\sqrt{\frac{f}{8}}$, in Eq. 2.100, it becomes

$$\frac{V}{V(f/8)^{1/2}} = 5.75\log_{10}\left(\frac{R}{k}\right) + 4.75 \tag{2.101}$$

By rearranging Eq. 2.101, we get

$$\frac{1}{\sqrt{f}} = 2.03\log_{10}\left(\frac{R}{k}\right) + 1.68 \tag{2.102}$$

Again on the basis of Nikuradse's experimental data for turbulent flow in rough pipe indicates that the experimental results follow closely the trend of the following equation instead of Eq. 2.102:

$$\frac{1}{\sqrt{f}} = 2.0\log_{10}\left(\frac{R}{k}\right) + 1.74 \tag{2.103}$$

In the case of turbulent flow in rough pipes, Eq. 2.103 is referred to as the Karman-Prandtl resistance equation.

Fig. 2.9 is obtained from the experimental results of by Nikuradse and shows the variation of friction factor with Reynolds number for smooth and rough pipes. It indicates the plot of f versus Re corresponding to different values of (R/k) on a logarithmic graph and is known as 'Stanton diagram'.

2.9.3 Hydrodynamically Smooth and Rough Boundary Criteria

The criteria of smooth, rough or in transition boundary is selected on the basis of the value of (k/δ') and it can be expressed as

$$\delta' = \frac{11.6v}{V_*} \tag{2.104}$$

Substituting $V_* = V\sqrt{\frac{f}{8}}$, in Eq. 2.104, it becomes

$$\delta' = \frac{11.6v}{V(f/8)^{1/2}} = \frac{11.6v(\sqrt{8})}{V\sqrt{f}} \tag{2.105}$$

Dividing Eq. 2.105 by R, we get

$$\frac{\delta'}{R} = \frac{11.6v(\sqrt{8})}{VR\sqrt{f}} = \frac{11.6\sqrt{8}(2v)}{VD\sqrt{f}} = \frac{65.6}{\text{Re}\sqrt{f}} \tag{2.106}$$

Eq. 2.106 shows the value of (δ'/R) is inversely proportional to the product of $(\text{Re}\sqrt{f})$. The value of (k/δ') is obtained by rearranging the Eq. 2.106 and can be expressed as

$$\frac{k}{\delta'} = \frac{\text{Re}\sqrt{f}}{65.6(R/k)} \tag{2.107}$$

The Eq. 2.107 represents that the nature of the boundary is also be defined by the parameter of (k/δ'), $\left(\frac{\text{Re}\sqrt{f}}{R/k}\right)$ and can be established by the Nikuradse's from the experimental data and it is indicated below

Rearranging the Eq. 2.98 and Eq. 2.103, it becomes

$$\frac{1}{\sqrt{f}} - 2.0\log_{10}(R/k) = 2.0\log_{10}\left(\frac{\text{Re}\sqrt{f}}{R/k}\right) - 0.8 \tag{2.108}$$

$$\frac{1}{\sqrt{f}} - 2.0\log_{10}\left(\frac{R}{k}\right) + 1.74 = 0 \tag{2.109}$$

From Eq. 2.108 and Eq. 2.109, it indicates that the smooth pipe $\left[\frac{1}{\sqrt{f}} - 2.0\log_{10}(R/k)\right]$ is a function of $\left(\frac{\text{Re}\sqrt{f}}{R/k}\right)$ while rough pipes $\left[\frac{1}{\sqrt{f}} - 2.0\log_{10}(R/k)\right]$ attains a constant value of 1.74. Based on experimental data of Nikuradse's a plot of $\left[\frac{1}{\sqrt{f}} - 2.0\log_{10}(R/k)\right] v/s\log_{10}\left(\frac{\text{Re}\sqrt{f}}{R/k}\right)$ is plotted and is shown in Fig. 2.10. From the plot it can be observed that as the pipe is hydrodynamically smooth when the parameter $\left(\frac{\text{Re}\sqrt{f}}{R/k}\right)$ is less than 17 and the pipe is hydrodynamically rough when parameter $\left(\frac{\text{Re}\sqrt{f}}{R/k}\right)$ is greater than 400. The pipe boundary is in transition state when the parameter $\left(\frac{\text{Re}\sqrt{f}}{Rk}\right)$ is lies between the value of 17 and 400. From Eq. 2.107, it is observed that the pipe is hydrodynamically smooth if (k/δ') is less than 0.25 and the pipe is hydrodynamically rough, if (k/δ') is greater than 6.0. The pipe boundary is in transition state when (k/δ') is lies between the value of 0.25 and 6.

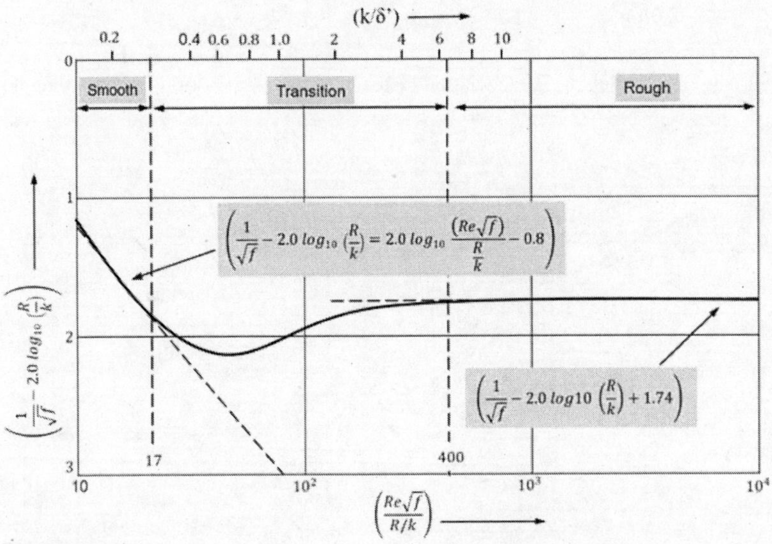

Fig. 2.10: Plot of $\left[\frac{1}{\sqrt{f}} - 2.0\log_{10}\left(\frac{R}{k}\right)\right]$ versus $\left[\log_{10}\left(\frac{Re/\sqrt{f}}{R/k}\right)\right]$ for artificially roughened pipes

2.9.4 Friction Factor Variation for Commercial Pipes

The various empirical relationships for the friction factor are developed from the experimental data of Nikuradse's for smooth, rough, and artificial roughness pipes and is plotted in Fig. 2.11 which cannot be used for calculating the friction factor for commercial pipes directly. Because commercial pipes have a very different wall roughness pattern than Nikuradse's uniform sand grain roughness. The number of experiments are performed at high Reynolds number in order to determine the equivalent sand grain diameter k for commercial pipes. The equivalent sand grain diameter k is computed by using the Eq. 2.110 as

$$\frac{1}{\sqrt{f}} = 2.0\log_{10}\left(\frac{R}{k}\right) + 1.74 \tag{2.110}$$

The value of friction factor (f) in Eq. 2.110 is calculated by using the Darcy-Welsbach equation which is derived in section 2.8..

In order to determine the friction factor in the transitional boundary for commercial pipes, the experimental test results of Nikuradse's and C.F. Colebrook are plotted in Fig. 2.11 by using the X ordinates as $\left(\frac{Re\sqrt{f}}{R/k}\right)$ and Y ordinates as $\left[\frac{1}{\sqrt{f}} - 2.0\log_{10}\left(\frac{R}{k}\right)\right]$ in log-log graph. From the graph, it is seen that the Colebrook and White have developed the empirical relationships to represent the variation of the friction factor for the commercial pipes and it is expressed as

$$\frac{1}{\sqrt{f}} - 2.0\log_{10}\left(\frac{R}{k}\right) = 1.74 - 2.0\log_{10}\left(1 + 18.7\frac{R/k}{\text{Re}\sqrt{f}}\right) \quad (2.111)$$

The Eq. 2.111 is used for calculating the friction factor in transitional region for commercial pipes.

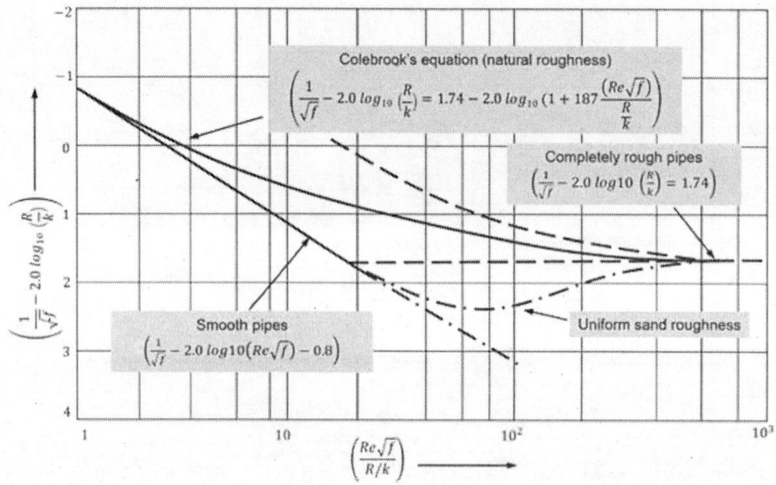

Fig. 2.11: Plot of $\left[\frac{1}{\sqrt{f}} - 2.0\log_{10}(R/k)\right]$ versus $\left[\log_{10}\left(\frac{Re\sqrt{f}}{R/k}\right)\right]$ for commercial pipes in transition region

EXAMPLE 2.12

Example 2.12 Determine the power required to pump 600 lps discharge of water through a 600 mm diameter pipe having total length of 5 km. The value of roughness height of pipe can be taken as 0.35 mm.

Solution

Given data:

$$D = 600 \text{ mm} = 0.60 \text{ m}$$

$$R = \frac{D}{2} = 0.3 \text{ m}$$

$$L = 5000 \text{ m}$$

$$Q = 600 \text{ litre/s} = 0.6 \text{ m}^3/\text{s}$$

$$k = 0.35 \text{ mm} = 0.35 \times 10^{-3} \text{ m}$$

For a rough pipe, the value of ' f ' is given by

$$\frac{1}{\sqrt{4f}} = 2\log_{10}(R/k) + 1.74$$

$$\frac{1}{\sqrt{4f}} = 2\log_{10}\left(\frac{.3}{.35 \times 10^{-3}}\right) + 1.74$$

$$\frac{1}{\sqrt{4f}} = 2\log_{10}(857.1) + 1.74$$

$$\frac{1}{\sqrt{4f}} = 7.606$$

$$\sqrt{4f} = \frac{1}{7.606}$$

$$f = (0.1314)^2/4 = 0.004316$$

The average velocity is given by

$$\mathrm{V} = \frac{\text{Discharge}}{\text{Area}}$$

$$\mathrm{V} = \frac{0.6}{\frac{\pi}{4}D^2}$$

$$\mathrm{V} = \frac{0.6}{\frac{\pi}{4}(.6)^2}$$

$$\mathrm{V} = 2.123 \text{ m/s}$$

Head lost due to friction is given by

$$h_f = \frac{4fLV^2}{d \times 2g}$$

$$h_f = \frac{4 \times .004316 \times 5000 \times 2.123^2}{0.6 \times 2 \times 9.81}$$

$$h_f = 33.049 \text{ m}$$

$$[\because V = 2.123, d = D = 0.6]$$

Power required is given by

$$P = \frac{\rho \times g \times Q \times h_f}{1000} \text{ kW}$$

$$P = \frac{1000 \times 9.81 \times 0.6 \times 33.049}{1000}$$

$P = 194.52$ kW. **Ans.**

EXAMPLE 2.12

Example 2.13 Find out the velocity for a fluid flow, where the pipe roughness at the surface just outside the laminar sub layer and at rough boundary of pipe will begin. The properties and the pipe specification area are as;

(i) Pipe diameter : 125 mm
(ii) Fluid : water at 15°C
(iii) Sand grain diameter : 0.70 mm
(iv) Kinematic viscosity of water at 15°C : 1.14×10^{-2} stokes.

EXAMPLE 2.13

Solution

Given data:

$$D = 0.125 \text{ m}$$

$$R = \frac{D}{2} = 0.0625 \text{ m}$$

$$k = 0.7 \times 10^{-3} \text{ m}$$

$$\nu = 1.14 \times 10^{-2} \text{ stokes}$$

$$\frac{R}{k} = \frac{0.0625}{0.70 \times 10^{-3}}$$

$$\frac{R}{k} = 89.285$$

For smooth pipe from Blasius equation, friction factor is given by

$$f = \frac{0.316}{\text{Re}^{1/4}}$$

Also, for smooth pipes, the friction factor is calculated by

$$\left[\frac{\text{Re}\sqrt{f}}{R/k}\right] = 17$$

$$(\text{Re}\sqrt{f}) = (17 \times 89.285)$$

Substituting value of f from the Blasius equation, it becomes

$$(\text{Re})^{\frac{7}{8}} = \frac{17 \times 89.285}{\sqrt{0.316}}$$

$$(\text{Re})^{\frac{7}{8}} = 2700.125$$

$$\text{Re} = 8347.98$$

The velocity at which the rough surface starts to disturb the laminar sublayer is

$$v = \frac{\text{Re} \times \nu}{D}$$

$$v = \frac{8347.98 \times 1.14 \times 10^{-2} \times 10^{-4}}{0.125}$$

$$v = 0.0761 \text{ m/s}$$

For rough pipes, the friction factor is calculated by

$$\sqrt{f} = \frac{1}{2\log_{10}\left(\frac{R}{k}\right) + 1.74}$$

$$\sqrt{f} = \frac{1}{2\log_{10}(89.285) + 1.74}$$

$$\sqrt{f} = 0.1772$$

EXAMPLE 2.13

In the limiting condition the parameter $\left[\frac{\text{Re}\sqrt{f}}{R/k}\right] = 400$

$$\text{Re} = \frac{400 \times 89.285}{0.1772}$$

$$\text{Re} = 2.01 \times 10^5$$

Thus, the minimum velocity at which a rough boundary will be formed by the pipe wall is

$$v = \frac{\text{Re}\ \nu}{D}$$

$$v = \frac{2.01 \times 10^5 \times 1.14 \times 10^{-2} \times 10^{-4}}{0.125}$$

$$v = 1.833 \text{ m/s.}$$ **Ans.**

EXAMPLE 2.13

Example 2.14 Determine the power loss through a pipe of length 1.5 km, if the pipe diameter is 500 mm and water is flowing with flow rate of 0.5 m^3/s. Take the wall thickness as 2.5 mm.

Solution:

Given data:

$$D = 500 \text{ mm} = 0.5 \text{ m}$$

$$R = \frac{0.5}{2} = 0.25 \text{ m}$$

$$Q = 0.5 \text{ m}^3/\text{s}$$

$$k = 2.5 \text{ mm} = 2.5 \times 10^{-3} \text{ m} = 0.0025 \text{ m}$$

$$L = 1.5 \text{ km} = 1500 \text{ m}$$

For rough pipes, the co-efficient of friction in terms of wall roughness, k is given by

$$\frac{1}{\sqrt{4f}} = 2\log_{10}(R/k) + 1.74$$

$$\frac{1}{\sqrt{4f}} = 2\log_{10}\left(\frac{0.25}{0.0025}\right) + 1.74$$

$$\frac{1}{\sqrt{4f}} = 5.74$$

$$\sqrt{4f} = \frac{1}{5.74}$$

$$\sqrt{4f} = 0.1742$$

$$4f = 0.03035$$

EXAMPLE 2.14

EXAMPLE 2.14

Also, the average velocity is given by

$$v = \frac{Q}{A}$$

$$v = \frac{0.5}{\frac{\pi}{4}(0.5^2)}$$

$$v = 2.547 \text{ m/s}$$

The head loss due to friction is given by,

$$h_f = \frac{4f \times L \times v^2}{D \times 2g}$$

$$h_f = \frac{0.03035 \times 1500 \times 2.547^2}{0.5 \times 2 \times 9.81}$$

$$h_f = 30.10 \text{ m}$$

The power lost is given by,

$$P = \frac{\rho g \times Q \times h_f}{1000}$$

$$P = \frac{1000 \times 9.81 \times 0.5 \times 30.10}{1000}$$

$P = 147.64$ kW. **Ans.**

EXAMPLE 2.15

Example 2.15 A 200 mm diameter pipe carrying a liquid (specific gravity: 0.90; absolute viscosity: 6.6×10^{-4} Ns/m^2) of discharge of 50 lps. What type (smooth or rough) of pipe it will be if a total head loss of 5.0 m occurs over a pipe length of 120 m.

Solution:

Given data:

D= 0.2 m

$$R = \frac{D}{2} = 0.1 \text{ m}$$

Specific gravity = 0.9

$\mu = 6.6 \times 10^{-4}$ Ns/m^2

$L = 120$ m

$h_f = 5$ m

Mean velocity, $V = \frac{Q}{A}$

$$V = \frac{50 \times 10^{-3}}{\frac{\pi}{4}(0.2)^2}$$

$$V = 1.591 \text{ m/s}$$

Reynolds number, is given by

$$Re = \frac{\rho VD}{\mu}$$

$$Re = \frac{0.90 \times 1000 \times 1.591 \times 0.20}{6.6 \times 10^{-4}}$$

$$Re = 4.339 \times 10^{5}$$

Hence fluid flow is turbulent.

Darcy-Weisbach equation is given by

$$h_f = \frac{fLV^2}{2gD}$$

By substituting the given values, we get

$$5 \quad = \frac{f \times 120 \times (1.591)^2}{2 \times 9.81 \times 0.2}$$

$$f = 0.06459$$

Suppose the pipe is hydrodynamically rough, then the friction factor is calculated as follows:

$$\frac{1}{\sqrt{f}} = 2.0\log_{10}\left(\frac{R}{k}\right) + 1.74$$

$$\frac{1}{\sqrt{0.06459}} = 2.0\log_{10}\left(\frac{R}{k}\right) + 1.74$$

$$(R/k) = 12.513$$

Thus

$$\left[\frac{Re\sqrt{f}}{\frac{R}{k}}\right] = \frac{4.339 \times 10^{5} \times (0.06459)^{\frac{1}{2}}}{12.513}$$

$$\left[\frac{Re\sqrt{f}}{\frac{R}{k}}\right] = 8812.737776 \approx 8813$$

Since the parameter $\left[\frac{Re\sqrt{f}}{R/k}\right]$ is greater than 17. Hence the pipe is hydro dynamically rough.

EXAMPLE 2.15

2.10 MOODY'S DIAGRAM

As discussed earlier, friction factor is the strong function of Reynolds number, considering this concept, L.F. Moody has plotted curves between friction factor and Reynolds number for different values of surface roughness to pipe diameter ratio (k/D) which is known as 'Moody's diagram' as shown in Fig. 2.12.

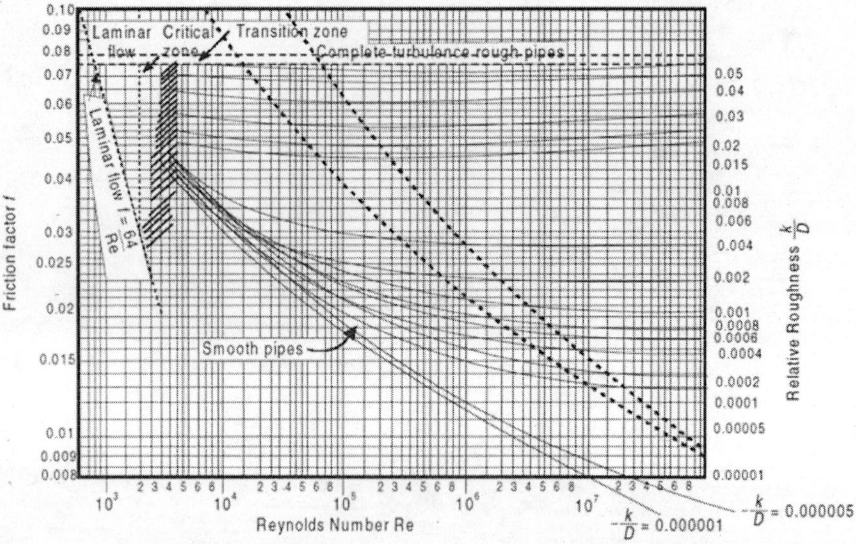

Fig. 2.12: Moody's diagram for friction factor for commercial pipes [2]

This diagram is mostly used for commercial pipes. The value of roughness height (k) for these pipes can be taken from Table 2.1. For a given pipe diameter (D) and selected value of k, the value of (k/D) is determined. The friction factor value can be determined from the Moody's diagram corresponding to the values of Reynolds number and roughness height to pipe diameter ratio(k/D).

The values of roughness height, (k) considered for different pipe material are given in Table 2.1.

Table: 2.1 Roughness height (k) for different pipe materials

S. No.	Pipe Materials	K in mm
1.	Glass, Brass, Copper, Lead	Smooth
2.	Steel, Wrought iron	0.045
3.	Asphalted cast iron	0.120
4.	Galvanized iron	0.150
5.	Cast iron	0.260
6.	Concrete	0.30 to 3.0
7.	Riveted steel	0.90 to 9.0

The roughness of the pipes increases with age as a result of corrosion. According to Colebrook and White, boundary roughness may increase with time approximately according to the following equation:

$$k = k_0 + \alpha t \tag{2.112}$$

Where,

k_0 is the equivalent roughness of a sand grain of new pipe material,

k is the equivalent roughness of a sand grain at any time t, and

α is the time rate of increase of roughness.

It will be possible to estimate the value of k at any future time by measuring resistance at two different times on the same pipe.

UNIT SUMMARY

- In order to have the exact solutions of fluid flow in open channel, the types of flow are required to be discussed as
  (i) Steady and unsteady flow (ii) uniform and non-uniform flow (iii) laminar and turbulent flow (iv) critical, sub-critical and super-critical flow

  laminar and turbulent flow

  $Re \leq 600$ for laminar flow and

  $Re \geq 2000$ for turbulent flow

  $600 > Re < 2000$ for transition state in general.

  critical, sub-critical and super-critical flow

  $Fr < 1$ for sub-critical which can also be called as tranquil or streaming flow

  $Fr = 1$ for critical

  $Fr > 1$ for super critical

- For Poisuielle flow – Viscous flow between two fixed parallel plates

  Velocity distribution, v

  $$v = \frac{1}{2\mu}\left(-\frac{\partial p}{\partial x}\right)(Dy - y^2)$$

  Discharge q per unit width

  $$q = \frac{D^3}{12\mu}\left(-\frac{\partial p}{\partial x}\right)$$

  Mean velocity, V

  $$V = \frac{2}{3}v_{max}$$

  Pressure drop $(p_1 - p_2)$ in the direction of flow

  $$(p_1 - p_2) = \frac{12\mu VL}{D^2}$$

  Maximum shear stress distribution, τ_{max}

$$\tau_{max} = \left(-\frac{\partial p}{\partial x}\right)\left(\frac{D}{2}\right)$$

- For Couette flow – Viscous flow between moving plate over fixed parallel plate

  Velocity distribution, v

$$v = \frac{V}{D}y - \frac{1}{2\mu}\left(\frac{\partial p}{\partial x}\right)(Dy - y^2)$$

  Shear stress distribution, τ

$$\tau = \mu\frac{V}{D} + \left(-\frac{\partial p}{\partial x}\right)\left(\frac{D}{2} - y\right)$$

- For laminar flow through in circular conduits

  Velocity distribution, v

$$v = \frac{1}{4\mu}\left(-\frac{\partial p}{\partial x}\right)(R^2 - r^2)$$

  Mean velocity of flow, V

$$V = \frac{1}{2}v_{\max}$$

- Hagen-Poiseuille equation

$$p_1 - p_2 = \frac{128\mu QL}{\pi D^4}$$

- Friction factor f for laminar flow in a pipe is given by

$$f = \frac{64}{\text{Re}}$$

- Shear velocity has dimensions of velocity and is given by

$$V_* = \sqrt{(\tau_0/\rho)}$$

$$V_* = V\sqrt{\frac{f}{8}}$$

- For laminar flow through annulus

  Mean velocity of flow

$$V = \frac{1}{8\mu}\left(-\frac{\partial p}{\partial x}\right)\left[R_1^2 + R_2^2 - \frac{(R_1^2 - R_2^2)}{\log_e(R_1/R_2)}\right]$$

- Concept of boundary layer

  The velocity increases from zero at the boundary surface to the velocity of the main stream asymptotically and thickness of this region is known as boundary layer thickness and represented by 'δ'.

- Measures of boundary layer thickness

  Displacement thickness $(\delta^*) = \int_0^{\delta}\left(1 - \frac{v}{V}\right)dy$

Momentum thickness $(\theta) = \int_0^\delta \frac{v}{V}\left(1 - \frac{v}{V}\right) dy$

Energy thickness $(\delta_E) = \int_0^\delta \frac{v}{V}\left(1 - \frac{v^2}{V^2}\right) dy$

- The loss of head due to friction in pipes is given by Darcy-Weisbach equation as

$$h_L = \frac{fLV^2}{2gD}$$

- For hydrodynamically smooth pipes the friction factor f is given by

Blasius formula

$$f = \frac{0.316}{(\mathrm{Re})^{1/4}} \text{ for } 4 \times 10^3 < \mathrm{Re} < 10^5$$

Karman-Prandlt resistance equation

$$\frac{1}{\sqrt{f}} = 2.0 \log_{10}\left(\mathrm{Re}\sqrt{f}\right) - 0.8 \text{for Re} > 10^5$$

Nikuradse's empirical equation

$$f = 0.0032 + \frac{0.021}{(\mathrm{Re})^{0.237}} \text{ for Re} > 10^5$$

- For hydrodynamically rough pipes the friction factor f is given by

Karman-Prandtl resistance equation

$$\frac{1}{\sqrt{f}} = 2.0 \log_{10}\left(\frac{R}{k}\right) + 1.74$$

For commercial pipes the friction factor f is given by

$$\frac{1}{\sqrt{f}} = -2.0 \log_{10}\left(\frac{R}{k}\right)$$

$$\frac{1}{\sqrt{f}} = 1.74 - 2.0 \log_{10}\left(1 + 18.7\frac{R/k}{\mathrm{Re}\sqrt{f}}\right)$$

EXERCISES

Multiple Choice Questions

2.1 The velocity in an incompressible Couette flow, how does the velocity vary?

a) Linear
b) Parabolic
c) No variation
d) Hyperbolic

2.2 Is the effect of increasing the thickness between two plates on the shear stress in case of a Couette flow?

a) Increases b)Decreases
c) Remains same d) Becomes infinite

2.3 In a Couette flow, where does the maximum temperature occur?
a) Top plate b) Bottom plate
c) Midpoint between the two plates d) No maximum temperature occurs

2.4 A pipe with a radius of 10 cm experiences laminar flow with an average velocity of 5 m/s. The velocity at 5 cm radius is
a) 7.5 m/s b) 10 m/s c) 2.5 m/s d) 5 m/s

2.5 In laminar flow loss of pressure head is proportional to;
a) Velocity b) Square of velocity
c) Cube of velocity d) Half of velocity

2.6 In case of a laminar flow through a long pipe of diameter d, the pressure head loss per unit length is
a) direct proportion to $\frac{\pi}{4} d^2$ b) directly proportion to d
c) In inverse proportion to $\frac{1}{\frac{\pi}{4}d^2}$ d) In inverse proportion to $\frac{1}{d}$

2.7 The ratio of maximum velocity to average velocity in a fully developed laminar viscous flow in a pipe is;
a) 3.0 b) 2.0 c) 2.5 d) 1.5

2.8 Which of the following parameter decides the flow in 'laminar' or 'turbulent'?
a) Reynold's number b) Mach number
c) Froude number d) Knudsen number

2.9 The statement of "The vertical distance from boundary surfaces in that velocity riches 99% in the free stream velocity, called as Boundary layer thickness" is;
a) True b) False

2.10 The boundary layer is formed as the when fluid comes in contact with solid surface boundary layers formed due to;
a) Surface tension b) Forces of adhesion
c) Force of gravity acting on the fluid d) Viscosity of the fluid

2.11 Out of the for which case a boundary is known as hydrodynamically smooth viscous form?
a) K/δ= 0.3 b) K/δ> 0.3 c) K/δ< 0.25 d) K/δ= 0.5

2.12 The Darcy friction factor for laminar flows is a consequence of-
a) Bernoulli's law b) Poiseuille's law c) Euler's law d) Newton's law

2.13 If 'f' is the Darcy-coefficient then head loss due to friction is represented by given as;
a) $\frac{fLV^2}{gd}$ b) $\frac{fLV^2}{2gd}$ c) $\frac{4fLV^2}{2gd}$ d) $\frac{16fLV^2}{2gd}$

2.14 If the flow velocity in a pipe is increased by 10% then the head loss due to friction increases by;

a) 21% b) 25% c) 5% d) 11%

2.15 The head loss due to friction in pipe reduces with;

a) Increased length of pipe b) Increased diameter of pipe

c) Decreased length of pipe d) Both b and c

2.16 For a steady laminar flow in a pipe of diameter D for given discharge, the head loss is proportion to;

a) D^{-1} b) D^{-2} c) D^{-3} d) D^{-4}

2.17 A pipe having 400 mm diameter and length of 1000 m carrying water at a velocity of 1.5 m/s. If friction factor is 0.0075 then head loss will be;

a) 20.2 m b) 12.8 m c) 35.6 m d) 6.4 m

2.18 In hydrodynamically smooth pipes, the friction factor depends on;

a) Reynolds number for laminar flow and relative smoothness for turbulent flow

b) Reynolds number for laminar flow and Reynolds number for turbulent flow

c) Relative smoothness for laminar flow and relative smoothness for turbulent flow

d) Relative smoothness for laminar flow and Reynolds number for turbulent flow

2.19 For smooth pipes, friction factor for turbulent flow is expressed in forms of Reynold's number Re as;

a) 0.64/R b) 64/R c) 0.31 $6/R^{1/4}$ d) $0.316/R^{4/5}$

2.20 The Moody's chart represents which parameter other than Darcy Weisbach friction factor?

a) Density of fluids b) Reynolds number

c) Viscosity of the fluid d) Slope of the inclination of the fluid

Answers of Multiple-Choice Questions

2.1 (a), 2.2 (b), 2.3 (c), 2.4 (a), 2.5 (a), 2.6 (d), 2.7 (b), 2.8 (a), 2.9 (a), 2.10 (d), 2.11 (c), 2.12 (b), 2.13 (c), 2.14 (a), 2.15 (d), 2.16 (d), 2.17 (b), 2.18 (b), 2.19 (c), 2.20 (b).

Short and Long Answer Type Questions

2.1 What is fluid dynamics.

2.2 Explain types of flow in order to have the exact solutions of fluid flow.

2.3 What is the significance of Reynolds number and Froude number?

2.4 Define velocity distribution in open channel

2.5 Define a Poisuielle flow and derive the velocity and shear stress distribution for viscous flow between two fixed parallel plates.

2.6 Define a Couette flow and derive the velocity and shear stress distribution for Couette flow.

2.7 Draw the velocity distribution and shear stress distribution profiles for Poisuielle flow and Couette flow.

2.8 Derive an expression for the velocity distribution for viscous flow through circular conduits. Show with help of figure, velocity distribution and shear stress in a circular pipe.

2.9 What is shear velocity and derive its expression

2.10 A viscous fluid flowing through an annulus derive expressions for: Discuss (i) Discharge through the annulus (ii)Average velocity of flow, and (iii) Shear stress distribution.

2.11 Discuss boundary layer theory and define boundary layer.

2.12 Define boundary layer thickness and displacement thickness with its expression.

2.13 What is momentum thickness in case of boundary layer and write an experience for it? Write an expression for the momentum thickness.

2.14 Define energy thickness. Write an expression for the energy thickness.

2.15 Derive the expression for head losses due to friction in a pipe (Darcy Weisbach equation)

2.16 Prove that the friction factor is inversely proportional to the Reynolds number in case of laminar flow in circular pipes.

2.17 Define hydrodynamically smooth and rough pipes.

2.18 Derive the Hagen-Poiseuille equation with suitable assumptions.

2.19 Considering Hagen-Poiseuille equation derive the head loss expression for flow in a pipe.

2.20 Discuss the flow through a rough pipe with respect to that in smooth pipe.

2.21 What is the significance of Moody's diagram?

Numerical Problems

2.22 A circular pipe with a diameter of 60 mm and a length of 500 m is carrying an oil of viscosity 0.9 poise and relative density 0.92. Flow rate is 0.0025 m^3/s. Calculate the pressure drop along a 250 m length as well as the shear stress at the wall.

2.23 An horizontal pipe with a diameter of 60 mm carrying glycerine exhibits a shear stress of 210 N/m^2 at its boundary. Find the pressure gradient, mean velocity, and Reynolds number. Take for glycerine r = 1280 kg/m^3 and u = 85.45 x 10-2 $N.s/m^3$

2.24 An oil having viscosity of 1.5 poise and specific gravity 0.92 flows through a pipe 30 mm diameter and 500 m long at 1/10 of the critical velocity for which Reynolds number is 2300. Find (a) the velocity of flow through the pipe; (b) the head in metres of oil across the pipe length required to maintain the flow; and (c) the power of the flow.

2.25 The maximum velocity of flow in a 150 mm diameter pipe is 2.5 m/s. For a laminar flow in a pipe, find: (i) The average velocity and its radius, and (ii) The velocity at 50 mm from the pipe wall.

2.26 An oil having dynamic viscosity μ = 1.80 Pa.s is flowing through a pipe of diameter of 0.20 m under laminar flow condition. If the flow velocity distribution is parabolic with maximum velocity of 2.5 m/s at centre of pipe. Find out the shearing stresses (i) at the pipe wall (ii) within the fluid at 20 mm from the wall.

2.27 Using the following data, determine the pressure drop in a pipe through which oil (viscosity = 0.1 Ns/m^2, relative density = 0.9) is flowing;

(i) Dia of pipe : 50 mm
(ii) Length of pipe : 450 m
(iii) Flow rate : 3.5 lps

2.28 In a 250 mm diameter pipe, determine the pressure gradient and average velocity of a fluid flowing through the pipe. Take the oil properties and other conditions as follows;

(i)	Viscosity	: 0.9 Ns/m^2
(ii)	Specific gravity	: 1.5
(iii)	Maximum shear stress at wall	: 300 N/m^2

2.29 For the velocity distribution in the boundary layer given by,

$$\frac{u}{U} = 3\left(\frac{y}{\delta}\right) - 2\left(\frac{y}{\delta}\right)^2$$

Determine the displacement thickness, momentum thickness, and energy thickness.

2.30 Determine the displacement, momentum, and energy thickness for a boundary layer of $\frac{u}{U} = \frac{3y}{2\delta}$ where u represents the velocity of the plate at a distance of y from it and $u = U$ at $y = \delta, \delta$ is the boundary layer thickness.

2.31 A velocity distribution in the boundary layer can be expressed as follows:

$$\frac{v}{V} = \frac{5}{2}\eta - \frac{3}{2}\eta^2$$

If $\eta = (y/\delta)$, determine (δ^*/δ) and (θ/δ).

2.32 Find out the maximum distance for air flowing over a flat plate at wind velocity of 150 km/h, if the laminar boundary layer exists up to Reynold number equal to 3 ×105. Use the standard value of kinematic viscosity of air.

2.33 The centre-line velocity of a 400 mm diameter pipe with turbulent flow is 5 m/s, and it is 4 m/s at 60 mm from the pipe wall. Calculate the shear friction velocity.

2.34 Water is flowing at a rate of 0.035 m^3/s in a smooth pipe of diameter 0.5 m and length 1000 m. Taking the kinematic viscosity of water to be 0.018 Stokes, find: (i) Head loss due to friction, (ii) Thickness of laminar sublayer (iii) Centre-line velocity, and (iv) Wall shear stress.

2.35 If a pipe has a diameter of 2.5 meters and is intended to carry an amount of water of 6.5 m^3/s with a minimum amount of energy expenditure, determine the height of the roughness projection on the wall of the pipe. ν for water is 0.017 x 10^{-4} m^2/s.

2.36 The flow rate of water in a rough pipe with a diameter of 350 mm and a length of 750 m is 0.35 m3/s. The wall roughness is 0.012 mm. Calculate the coefficient of friction, wall shear stress, centre line velocity, and velocity at 150 mm from the pipe wall.

2.37 Determine the power required to pump 500 lps discharge of water through a 550 mm diameter pipe having total length of 7 km. The value of roughness height of pipe can be taken as 0.25 mm.

2.38 Determine the power loss through a pipe of length 2 km, if the pipe diameter is 400 mm and water is flowing with flow rate of 0.45 m^3/s. Take the wall thickness as 2 mm.

2.39 A 150 mm diameter pipe carrying a liquid (specific gravity: 0.92; absolute viscosity: 6 × 10-4 Ns/m^2) of discharge of 55 lps. What type (smooth or rough) of pipe it will be if a total head loss of 10 m occurs over a pipe length of 150 m.

2.40 Find out the velocity for a fluid flow, where the pipe roughness at the surface just outside the laminar sub layer and at rough boundary of pipe will begin. The properties and the pipe specification are as;

(i) Pipe diameter : 250 mm
(ii) Fluid : water at 15°C
(iii) Sand grain diameter : 0.65 mm
(iv) Kinematic viscosity of water at 15°C : 1.2×10^{-2} stoke

2.41 An oil (kinematic viscosity: 0.19 stokes) is to be carried through a pipe having roughness as 0.035 mm. Determine the pipe diameter so that the head loss should not exceed 0.55 m. following data be used in calculation.

(i) Friction factor : $f = 0.0055\left[1 + \left(20000\frac{\varepsilon}{D} + \frac{10^6}{Re}\right)^{0.33}\right].$
(ii) Flow velocity : 2 m/s
(iii) Length of pipe : 800 m

2.42 Find out the diameter of a pipe for the following data;

(i) Discharge : $0.15\ m^3/s$
(ii) Length of pipe : 250 m
(iii) Head loss : 10 m
(iv) Kinematic viscosity of water at 20° C : $1.2\times10^{-6}\ m^2/s$
(v) Roughness height : 0.15 mm

2.43 Find out the head losses in a pipe and power required to pump the water for the following data;

(i) Length of pipe : 600 m
(ii) Pipe diameter : 150 mm
(iii) Discharge : 55 lps
(iv) Roughness height : 0.2 mm
(v) Kinematic viscosity : $1.15\times 10^{-6}\ m^2/s$

REFERENCES AND SUGGESTED READINGS

List of some of the books is given below which may be used for further learning of the subject:

1. Yunus A. Çengel, Fluid Mechanics: Fundamentals and Applications, McGRAW-HILL publication, 2006.
2. P. N. Modi, Hydraulics & Fluid Mechanics including Hydraulics Machines, Rajsons Publications Pvt. Ltd., 2014.
3. R. K. Bansal., "A Textbook of Fluid Mechanics and Hydraulic Machines", Laxmi Publications Pvt. Ltd. 2010.
4. Er. R.K. Rajput., "Fluid Mechanics and Hydraulic Machines", S. Chand & Company Ltd. 2011.
5. S. K. SOM and G Biswas., "Introduction to Fluid Mechanics and Fluid Machines", Tata McGraw-Hill Publishing Company Limited New Delhi 2008.
6. Yunus A. Çengel and John M. Cimbala., "Fluid Mechanics: Fundamentals and Applications", McGraw Hill Publishing. 2006.

3 Dimensional Analysis

UNIT SPECIFICS

Through this unit we have discussed the following aspects:

- *Need for dimensional analysis;*
- *Methods of dimensional analysis;*
- *Similitude and its types;*
- *Dimensionless parameters;*
- *Problems including application of dimensionless parameters;*
- *Model analysis.*

The practical applications of the topics are discussed for generating further curiosity and creativity as well as improving problem solving capacity.

Besides giving a large number of multiple-choice questions as well as questions of short and long answer types marked in two categories following lower and higher order of Bloom's taxonomy, assignments through a number of numerical problems, a list of references and suggested readings are given in the unit so that one can go through them for practice. It is important to note that for getting more information on various topics of interest some QR codes have been provided in different sections which can be scanned for relevant supportive knowledge.

RATIONALE

This fundamental unit on dimensional analysis helps students to deal the fluid mechanics related problems applying the dimensional and non-dimensional concepts of physical quantities. Using similitude has been presented for applying similarity law which includes geometric, kinematic and dynamic similarity. Further, non-dimensional quantities are also presented which will help the students in analysing the fluid flow under different conditions. All these are discussed at length to develop the subject. Some related problems are pointed out with an extension to practical application of dimensional and non-dimensional parameters which can help further for getting a clear idea of the concern topics on fluid mechanics.

Dimensional analysis is an important technique to analyze the fluid flow problems through dealing the relationship between the physical quantities. It deals to facilitate the establishment of the relationship between the physical quantities in terms of dimensions. Its practical applications are related to the homogeneity testing of fluid motion expressions, establishing expressions for non-dimensional parameters with respect to their relative significance, systematic expressing of experimental results obtained from model testing in terms of non-dimensional parameters to analyse of complex fluid flow phenomenon.

PRE-REQUISITES

Mathematics: Algebraic equations (Class XII)

Physics: Mechanics (Class XII)

UNIT OUTCOMES

List of outcomes of this unit is as follows:

U3-O1: Describe the need for dimensional analysis

U3-O2: Describe the methods of dimensional analysis

U3-O3: Explain similitude and its types

U3-O4: Realize the role of dimensionless parameters in practical applications

U3-O5: Describe the model analysis and solving dimensional analysis problems

Unit-3 Outcomes	EXPECTED MAPPING WITH COURSE OUTCOMES (1- Weak Correlation; 2- Medium correlation; 3- Strong Correlation)					
	CO-1	CO-2	CO-3	CO-4	CO-5	CO-6
U3-O1	3	3	3	-	3	1
U3-O2	1	1	2	2	-	1
U3-O3	2	1	3	1	2	1
U3-O4	-	-	2	1	3	2
U3-O5	3	2	3	-	2	1

3.1 DIMENSIONAL ANALYSIS

Science and engineering problems are solved by knowing the physical quantities which are expressed in terms of physical dimensions. Dimensional analysis deals with the relationship between the physical quantities which are analysed with the help of dimensions and units of measurements. Any physical phenomenon is expressed in terms of physical quantities either in dimensional form or non- dimensional forms. 'Dimensional Analysis' deals with to facilitate the establishment of the relationship between the physical quantities in terms of dimensions. Further, by combining dimensional variables, non-dimensional parameters can be formed. In general, problems of fluid mechanics such as; dimensional homogeneity testing of fluid motion expressions, establishing expressions for non-dimensional parameters with respect to their relative significance and systematic expressing of experimental results obtained from model testing in terms of non-dimensional parameters to analyse of complex fluid flow phenomenon.

The physical quantities can be classified as primary quantities and secondary quantities. The primary quantities also known as fundamental quantities which are independent on each other such as; mass, length, time, and temperature. These are respectively represented by M, L, T, θ and are useful for particular cases. The secondary quantities are derived from primary quantities and also known as derived quantities. The examples of secondary quantities are; area, volume, velocity, acceleration, force, energy, power etc. The dimensions of physical quantities are expressed in terms of the derived quantities expressions. To have more clarity, let us consider a secondary quantity of force.

We know that force can be defined as;

$$\text{Force} = \text{Mass} \times \text{Acceleration}$$

Where, mass and acceleration can be expressed in terms of primary quantities as;

$$[\text{Mass}] = [\text{M}]$$

$$[\text{Accelration}] = \left[\frac{\text{L}}{\text{T}^2}\right]$$

Thus, force can be written in terms of primary quantities as;

$$[\text{Force}\ ] = [\text{M}] \times \left[\frac{\text{L}}{\text{T}^2}\right] = [\text{MLT}^{-2}] \tag{3.1}$$

The form of dimension of a given quantity is independent of the unit system (SI, Metric, or English), which facilitates to convert the units from one measurement system to the other.

3.2 METHODS OF DIMENSIONAL ANALYSIS

In several fluid mechanics problems, the relationship between dimension variables of physical system can be established by two methods such as Rayleigh's method and Buckingham's π-theorem method which are discussed as follows.

3.2.1 Rayleigh's Method

If number of independent variables are maximum three or four, Rayleigh's method can be used to find the expression for dependent variables. However, it is difficult to determine the expression for a variable if number of variables are more than four by this method.

Mathematically, Rayleigh's method can be expressed as follows, if a dependent variable, X depends on variables X_1, X_2 and X_3.

$$X = f[X_1, X_2, X_3] \tag{3.2}$$

Further, the above equation 3.2 is also written as

$$X = CX_1^a \cdot X_2^b \cdot X_3{}^c \tag{3.3}$$

where C is constant and a, b and c are arbitrary powers.

Now, the values for a, b and c are determined by relating the powers of the fundamental dimension on left- and right-hand sides.

Accordingly, the final expression can be obtained for dependent variable (X).

To understand this method further following solved examples are given.

SCAN ME

For better understanding dimensional analysis

Example 3.1 Establish an expression to represent the time period for a pendulum, considering the following parameters

(i) time period of pendulum: T

(ii) length of pendulum: L

(iii) acceleration due to gravity: g

Solution

Basically, time period depends on length and acceleration due to gravity which can be expressed as $T \propto (L, g)$

As per the dimensional analysis it can be written as

$T = CL^a \cdot g^b$

where C is a constant and g can be expressed in primary dimensions as LT^{-2}

$T^1 = CL^a.(LT^{-2})^b$

Putting the dimensions as equal on both sides,

Power of T; $1 = -2b$ or $b = -1/2$

Power of L; $0 = a + b$ or $a = -b$ or $1/2$

Putting the values of a and b in time period expression, thus

$$T = CL^{1/2} \cdot g^{-1/2}$$

$$T = C\sqrt{\frac{L}{g}}$$

Based on experimental study, the value of C is determined as 2π

It is therefore, the time period can be expressed as

$T = 2\pi\sqrt{\frac{L}{g}}$. **Ans.**

EXAMPLE 3.1

Example 3.2 A sphere having diameter D, moving in a fluid with a velocity of v. Derive the expression for drag force if the viscosity and density of fluid are represented by μ and ρ respectively.

Solution

We know that drag force F is a function of diameter (D), velocity (v), density (ρ) and viscosity (μ). Mathematically, it can be expressed as

$$F \propto (D, v, \rho, \mu)$$

$$F = C \cdot D^a . v^b \cdot \rho^c \cdot \mu^d$$

where C is a constant.

Considering the primary dimensions for each parameter and it can be expressed as

$$MLT^{-2} = CL^a \cdot (LT^{-1})^b \cdot (ML^{-3})^c \cdot (ML^{-1}T^{-1})^d$$

Putting the dimensions as equal on both sides

Power of M; $1 = c + d$

Power of L; $1 = a + b - 3c - d$

Power of T; $-2 = -b - d$

As there are four unknowns with only three equations, therefore three variables can be expressed in terms of fourth variable. Here, the most important parameter is considered as density and three unknowns (a, b, c) are expressed in terms of d i.e., the power of viscosity therefore,

$$c = 1 - d$$

$$b = 2 - d$$

$$a = 1 - b + 3c + d$$

or,

$$a = 1 - 2 + d + 3(1 - d) + d$$

or,

$$a = 1 - 2 + d + 3 - 3d + d$$

or,

$$a = 2 - d$$

Putting the values of a, b, c, d in force equation, thus

$$F = KD^{2-d} \cdot v^{2-d} \cdot \rho^{1-d} \cdot \mu^d$$

$$= KD^2 v^2 \rho (D^{-d} \cdot v^{-d} \cdot \rho^{-d} \cdot \mu^d)$$

$$= K\rho D^2 v^2 \left(\frac{\mu}{\rho v D}\right)^d$$

$= K\rho D^2 v^2 \phi\left(\frac{\mu}{\rho v D}\right)$. **Ans.**

EXAMPLE 3.2

Example 3.3 The power of a pump (P) is the function of discharge (Q), specific weight of fluid and head (H). Derive the expression of power.

Solution

Mathematically, power can be expressed as

$$P \propto (w, Q, H)$$

$$P = C \cdot w^a . Q^b \cdot H^c$$

where C is a constant.

Considering the primary dimensions for each parameter in power equation and it can be expressed as

$$ML^2T^{-3} = C.(ML^{-2}T^{-2})^a.(L^3T^{-1})^b.L^c$$

Putting the dimensions as equal on both sides

Power of M; $1 = a$

Power of L; $2 = -2a + 3b + c$

Power of T; $-3 = -2a - b$

From expression power of T,

$$-3 = -2\,(1) - b$$

or, $b = 1$

From expression power of L,

$$2 = -2\,(1) + 3(1) + c$$

or, $c = 1$

Putting the values of a, b and c in power equation, thus

$$P = C \cdot w^1 . Q^1 \cdot H^1$$

$$P = CwQH$$

or, $P = C\rho gQH$. **Ans.**

EXAMPLE 3.3

Example 3.4 Prove that the resistance to the motion of a sphere moving in a fluid is represented as

$$R = (\rho D^2\, v^2)\phi\left(\frac{\mu}{\rho v D}\right)$$

Where, ρ is density, D is sphere of diameter, v is velocity and μ is viscosity. What will be the expression of resistance R in the same case for viscous flow for low velocity.

Solution

Resistance R is a function of

EXAMPLE 3.4

$$R \propto (\mu, \rho, D, v)$$
$$R = C(\mu^a.\rho^b.D^c.v^d)$$

where C is a constant.

Substituting primary dimensions for each parameter

$$MLT^{-2} = C(ML^{-1}T^{-1})^a \cdot (ML^{-3})^b \cdot L^c \cdot (LT^{-1})^d$$

Equalizing the dimensions of M, L and T on both sides

Power of M; $1 = a + b$

Power of L; $1 = -a - 3b + c + d$

Power of T; $-2 = -a - d$

Solving the above equations for b, c and d in terms of a

$$b = 1 - a$$
$$d = 2 - a$$
$$c = 1 + \mathrm{a} + 3b - d$$

or,

$$c = 1 + a + 3(1 - a) - 2 + a$$

or,

$$c = 1 + a + 3 - 3a - 2 + a$$

or,

$$\mathrm{c} = 2 - a$$

Substituting the values of a, b, c, d in resistance equation,

$$R = C\mu^a \cdot \rho^{1-a} \cdot D^{2-a} \cdot v^{2-a}.$$
$$R = C\rho D^2 v^2 (\mu^a \cdot \rho^{-a} \cdot D^{-a} \cdot v^{-a}.)$$
$$R = C\rho D^2 v^2 \left(\frac{\mu}{\rho v D}\right)^a$$
$$R = C\rho D^2 v^2 \phi\left(\frac{\mu}{\rho v \mathrm{D}}\right) \textbf{ Ans.}$$

EXAMPLE 3.4

Example 3.5: For a case given in Example 3.4 find out the viscosity for a liquid having specific gravity as 0. 85 and the diameter of the sphere is 0.9 mm and specific gravity is 7.8. Assume the flow velocity as 18 mm/sec and value of C as 3π.

Solution

Given data:

Diameter, $D = 0.9$ mm

EXAMPLE 3.5

Velocity, $v = 18$ mm/s

$$C = 3\pi$$

$$R = C\rho D^2 v^2 \phi\left(\frac{\mu}{\rho v D}\right) \text{ (From Example 3.4)}$$

$$R = C\mu v D$$

Resistance R is equal to gravitational force for uniform velocity of the sphere.

$$3\pi \times \mu \times 18 \times 10^{-3} \times 0.9 \times 10^{-3}$$

$$= \frac{4}{3}\pi\left(\frac{0.9}{2} \times 10^{-3}\right)^3 \times (7.8 - 0.85) \times 9810$$

Viscosity for a liquid is given by

$$\mu = \frac{4(0.45)^3(6.95 \times 9.81)}{3 \times 3 \times 18}$$

$$= 0.1534 \text{ N.s}/\text{m}^2$$

$\mu = 1.534$ poise. **Ans.**

EXAMPLE 3.5

3.2.2 Buckingham's π-Theorem

Buckingham's π-theorem minimizes the drawback faced in Rayleigh's Method having limitation of four number of variables. Number of sets of non-dimensional parameters having variables can be computed by Buckingham π-theorem without knowing the form of equation. For a case of, where n variables having m physical dimensions, the variables can be arranged in non-dimensional terms as (n-m) known as π -terms.

Followings steps will be useful to understand the Buckingham π-theorem. Let us assume X_1 is the dependent variable while other variables $X_2, X_3, \ldots, X_n$ are independent, then X_1 can be written as

$$X_1 = f(X_2, X_3, \ldots, X_n) \quad (3.4)$$

Further, it can be expressed as

$$f_1(X_1, X_2, X_3, \ldots, X_n) = 0 \quad (3.5)$$

The above expression can be written in terms of Buckingham π-theorem with m dimensional variables as

$$f(\pi_1, \pi_2, \ldots, \pi_{n-m}) = 0 \quad (3.6)$$

Where, π_1, π_2,…. are number of dimensionless groups known as π – terms.

Here all π-terms are independent and dimensionless. (m + 1) variables are the parts of π-terms.

Considering the above case having repeating variables as X_2, X_3 and X_4 we can write π – terms as

$$\begin{bmatrix} \pi_1 & = X_2^{a_1} \cdot X_3^{b_1} \cdot X_4^{c_1} \cdot X_1 \\ \pi_2 & = X_2^{a_2} \cdot X_3^{b_2} \cdot X_4^{c_2} \cdot X_5 \\ \vdots & \\ \pi_{n-m} & = X_2^{a_{n-m}} \cdot X_3^{b_{n-m}} \cdot X_4^{c_{n-m}} \cdot X_n \end{bmatrix} \quad (3.7)$$

The above equations can be solved by applying dimensional homogeneity principle and considering values of a_1, b_1, c_1 …. By putting these values in Eq. 3.7, the values of π-terms can be determined. Hence the final expressions will become as

$$\pi_1 = \phi[\pi_2, \pi_3, \dots, \pi_{n-m}] \quad (3.8)$$

$$\pi_2 = \phi_1[\pi_1, \pi_3, \dots, \pi_{n-m}] \quad (3.9)$$

Fluid systems have geometric properties, fluid properties and fluid flow properties. Buckingham π-theorem can be applied to solve the problems after proper selection of fluid mechanics related variables i.e., dependent variables and repeating variables. To select repeating variables, one variable should include one of the properties mentioned above. Further dependent variable should not be repeating variable. The classification of fluid mechanics related variables is given in Table 3.1.

Table 3.1: Classification of variables

Property	Variable
Geometric property	Length (L), Diameter (D), Height (H) etc.
Flow property	Velocity (v), acceleration (a) etc.
Fluid flow property	Viscosity (μ) , density (ρ) etc.

For selecting repeating variables following conditions are considered.

(i) Repeating variables should form a group other than dimensionless group
(ii) The number of repeating variables should be equal to fundamental dimensions
(iii) Each repeating variable should have different dimensions

To give more clarity about the application of Buckingham π-theorem some examples are given below.

EXAMPLE 3.6

Example 3.6: Using π-theorem, derive the expression for specific speed of a pump as

$$N_s = \frac{N\sqrt{Q}}{H^{3/4}}$$

Where, variable parameters are as discharge Q, speed N, head H, Impeller diameter D of the pump, density ρ of the fluid and acceleration due to gravity g.

Solution

Specific speed of a pump is function of

$$N = f(\rho, gH, Q, D)$$

In this case, *gH*, known as shaft work, is considered to be one variable only.

Total number of variables (n)=5

Number of fundamental dimensions (m)=3

ρ, gH and Q are selected as repeating variables.

Number of π-terms is given by

$$\pi \text{ terms} = n - m$$

$$\pi \text{ terms} = 5 - 3 = 2$$

$$\pi_1 = \rho^{a_1}(gH)^{b_1}Q^{c_1}N$$

and

$$\pi_2 = \rho^{a_2}(gH)^{b_2}Q^{c_2}D$$

Solving for π_1 terms

$$\pi_1 = \rho^{a_1}(gH)^{b_1}Q^{c_1}N$$

Substituting fundamental dimensions,

$$M^0L^0T^0 = (ML^{-3})^{a_1}.(L^2T^{-2})^{b_1}.(L^3T^{-1})^{c_1}.T^{-1}$$

Equalizing dimensions on both sides,

Power of M; $0 = a_1$

Power of L; $0 = -3a_1 + 2b_1 + 3c_1$

Power of T; $0 = -2b_1 - c_1 - 1$

Solving for a_1, b_1 and c_1

$$a_1 = 0, b_1 = -\frac{3}{4}, c_1 = \frac{1}{2}$$

Putting the value of a_1, b_1 and c_1 in π_1 terms

$$\pi_1 = \left[\frac{NQ^{1/2}}{(gH)^{3/4}}\right]$$

Solving for π_2 terms

$$\pi_2 = \rho^{a_2}(gH)^{b_2}Q^{c_2}D$$

Substituting fundamental dimensions,

$$M^0L^0T^0 = (ML^{-3})^{a_2}.(L^2T^{-2})^{b_2}.(L^3T^{-1})^{c_2}.\mathrm{L}$$

Equalizing dimensions on both sides,

Power of M; $0 = a_2$

Power of L; $0 = -3a_2 + 2b_2 + 3c_2 + 1$

Power of T; $0 = -2b_2 - c_2$

Solving for a_2, b_2 and c_2

$$a_2 = 0, b_2 = \frac{1}{4}, c_2 = -\frac{1}{2}$$

EXAMPLE 3.6

EXAMPLE 3.6

Putting the value of a_2, b_2 and c_2 in π_2 terms

$$\pi_2 = \left[\frac{D(gH)^{1/4}}{Q^{1/2}}\right]$$

It is therefore, final expression can be expressed as

$$\left[\frac{NQ^{1/2}}{(gH)^{3/4}}\right] = \phi\left[\frac{D(gH)^{1/4}}{Q^{1/2}}\right]$$

Substituting $Q = 1, H = 1$ and $N = N_s$ in the above equation, we get

$$N_s = \frac{N\sqrt{Q}}{H^{3/4}}.$$ **Ans.**

EXAMPLE 3.7

Example 3.7 Using dimensional analysis, show the torque due to friction developed over a disc rotating in a fluid is expressed as

$$T = D^5 N^2 \rho\phi\left[\frac{\mu}{D^2 N\rho}\right]$$

Where, D is diameter of disc, N is speed, μ is the viscosity and ρ is the density of fluid.

Solution

Torque due to friction is function of

$$T = f(D, N, \mu, \rho)$$

Total number of variables (n)=5

Number of fundamental dimensions (m)=3

D, N, ρ are selected as repeating variables.

Number of π-terms is given by

$$\pi \text{ terms} = n - m$$

$$\pi \text{ terms} = 5 - 3 = 2$$

$$\pi_1 = D^{a_1} \cdot N^{b_1} \cdot \rho^{c_1} \cdot T$$

and

$$\pi_2 = D^{a_2} \cdot N^{b_2} \cdot \rho^{c_2} \cdot \mu$$

Solving for π_1 terms

$$\pi_1 = D^{a_1} \cdot N^{b_1} \cdot \rho^{c_1} \cdot T$$

Substituting fundamental dimensions,

$$M^0L^0T^0 = L^{a_1} \cdot (T^{-1})^{b_1} \cdot (ML^{-3})^{c_1} \cdot ML^2T^{-2}.$$

Equalizing dimensions on both sides,

Power of M; $0 = c_1 + 1$

Power of L; $0 = a_1 - 3c_1 + 2$

Power of T; $0 = -b_1 - 2$

Solving for a_1, b_1 and c_1

$$a_1 = -5, b_1 = -2, c_1 = -1$$

Putting the value of a_1, b_1 and c_1 in π_1 terms

$$\pi_1 = \frac{T}{D^5N^2\rho}$$

Solving for π_2 terms

$$\pi_2 = D^{a_2} \cdot N^{b_2} \cdot \rho^{c_2} \cdot \mu$$

Substituting fundamental dimensions,

$$M^0L^0T^0 = L^{a_2} \cdot (T^{-1})^{b_2} \cdot (ML^{-3})^{c_2} \cdot ML^{-1}T^{-1}.$$

Equalizing dimensions on both sides,

Power of M; $0 = c_2 + 1$

Power of L; $0 = a_2 - 3c_2 - 1$

Power of T; $0 = -b_2 - 1$

Solving for a_2, b_2 and c_2

$$a_2 = -2, b_2 = -1, c_2 = -1$$

Putting the value of a_2, b_2 and c_2 in π_2 terms

$$\pi_2 = \frac{\mu}{D^2N\rho}$$

It is therefore, final expression can be expressed as

$$\frac{T}{D^5N^2\rho} = \phi\left(\frac{\mu}{D^2N\rho}\right)$$

or,

$$T = D^5N^2\rho\phi\left[\frac{\mu}{D^2N\rho}\right].$$ **Ans.**

EXAMPLE 3.7

3.3 SIMILITUDE

In some cases, analysis of fluid problems become very difficult in their actual system (prototype) sizes. In these situations, a reduced size of similar system can be analyzed in laboratory. The reduced size system is known as 'model'. The model and prototype can be said similar if these are: geometrically similar, kinematically similar and dynamically similar. This is known as 'Similarity law' under similitude. The similitude describes that model and prototype will be completely similar if they follow similarity law. In case of fluid machines, the performance of a prototype machine can be predicted from the results obtained through testing of model which is sized based on similarity law. The three types of similarity laws are discussed below.

3.3.1 Geometric Similarity

In case of geometric similarity, the ratio of two physical dimensions of an actual (prototype) is equal to the ratio of corresponding dimensions of model.

Mathematically, it can be expressed considering subscripts m and a for model and actual (prototype) respectively as

For length: $$\frac{L_m}{L_a} = \frac{B_m}{B_a} = \frac{D_m}{D_a} = l_r \qquad (3.10)$$

For area: $$\frac{A_m}{A_a} = \frac{L_m \times B_m}{L_a \times B_a} = l_r^2 \qquad (3.11)$$

For volume: $$\frac{V_m}{V_a} = \frac{L_m \times B_m \times D_m}{L_a \times B_a \times D_a} = l_r^3 \qquad (3.12)$$

SCAN ME

For better understanding similitude

3.3.2 Kinematic Similarity

In fluid mechanics, kinematics deals with the space-time relationship such as velocity, acceleration and flow rate of fluids. Kinematic similarity says that the ratio of two velocities of fluid flow through an actual (prototype) system is equal to the ratio of corresponding velocity through model.

Mathematically, it can be expressed as

$$\frac{(v_1)_m}{(v_2)_m} = \frac{(v_1)_a}{(v_2)_a} = k_v \qquad (3.13)$$

Similarly, for acceleration

$$\frac{(a_1)_m}{(a_2)_m} = \frac{(a_1)_a}{(a_2)_a} = k_a \qquad (3.14)$$

For flow rate

$$\frac{(Q_1)_m}{(Q_2)_m} = \frac{(Q_1)_a}{(Q_2)_a} = k_Q \qquad (3.15)$$

Kinematic similarity can also be expressed in scale ratios as follows

For velocity

$$v_r = \frac{v_m}{v_a} \qquad (3.16)$$

or, $$v_r = \left(\frac{L_m}{T_m} \Big/ \frac{L_a}{T_a} \right)$$

or, $$v_r = \frac{l_r}{T_r} \qquad \left(T_r = \frac{T_m}{T_a}\right) \qquad (3.17)$$

Similarly, for acceleration

$$a_r = \frac{a_m}{a_a} = \frac{l_r}{T_r^2} \qquad (3.15)$$

For flow rate

$$Q_r = \frac{Q_m}{Q_p} = \frac{l_r{}^3}{T_r} \qquad (3.15)$$

3.3.3 Dynamic Similarity

As per the dynamic similarity the ratio of two forces of fluid through a model is equal to the ratio of corresponding forces through prototype. In general, inertia forces (F_i), viscous forces (F_v), forces due to gravity (F_g), pressure forces (F_p), elastic forces (F_e) and forces due to surface tension (F_s) are the forces acting on a fluid. The inertia force is the product of mass (M) and acceleration of fluid (a) and as per the Newton's second law of motion, the sum of other forces will be equal to inertia force.

Mathematically, it can be expressed as

$$\Sigma F = F_v + F_g + F_p + F_e + F_s = F_i = Ma \tag{3.16}$$

Now, as per the dynamic similarity these forces of fluid over a model and prototype can be written as

$$\frac{(\Sigma\ F)_m}{(\Sigma\ F)_a} = \frac{(Ma)_m}{(Ma)_a} \tag{3.17}$$

The above expression includes all the forces and the individual components of each forces as per the complete dynamic similarity may be written as

For viscous force:

$$\frac{(Ma)_m}{(Ma)_a} = \frac{(F_v)_m}{(F_v)_a} \tag{3.17}$$

For gravity force:

$$\frac{(Ma)_m}{(Ma)_a} = \frac{(F_g)_m}{(F_g)_a} \tag{3.18}$$

For pressure force:

$$\frac{(Ma)_m}{(Ma)_a} = \frac{(F_p)_m}{(F_p)_a} \tag{3.19}$$

For elastic force:

$$\frac{(Ma)_m}{(Ma)_a} = \frac{(F_e)_m}{(F_e)_a} \tag{3.20}$$

For surface tension:

$$\frac{(Ma)_m}{(Ma)_a} = \frac{(F_s)_m}{(F_s)_a} \tag{3.21}$$

For dynamic similarity of the systems it is necessary that these should follow the geometrical similarity and kinematic similarity. In conclusion, similitude exist between actual system and its model if the systems follow all the three similarity laws.

3.4 DIMENSIONLESS PARAMETERS

Dimensionless parameter is defined as the quantity which does not have any physical dimensions. The dimensionless quantities are used in different fields. In fluid mechanics following dimensionless parameters known as dimensionless numbers are frequently used.

(i) Reynolds number

(ii) Froude number

(iii) Euler number

(iv) Weber number

(v) Mach number

Reynolds number (*Re*):

Reynolds number is the ratio of inertia force (F_i) to the viscous force (F_v) of flowing fluid.

$$Re = \frac{F_i}{F_v} \tag{3.22}$$

Mathematically, it can be derived as;

Inertia force, $F_i = m \times a$

$m = \rho \times V;\ a = \frac{v}{t}$

$$F_i = \rho \times v \times \frac{v}{t} \tag{3.23}$$

The Eq. 3.23 can be rearranged as

$F_i = \rho \times \frac{v}{t} \times v = \rho \times A \times v \times v$

$$F_i = \rho A v^2 \tag{3.24}$$

Similarly, for viscous force, $F_v = \tau \times A$

$$\tau = \mu \frac{du}{dy} = \mu \frac{v}{L} \tag{3.25}$$

Then

$$F_v = \mu \frac{v}{L} \times A \tag{3.26}$$

Where, m is mass, a is acceleration of flowing fluid, V is volume, v is velocity, A is area, t is time, τ is shear stress and L is length

Putting Eq. 3.24 and 3.26 in Eq. 3.22.

Thus,

$$Re = \frac{\rho A v^2}{\mu \frac{v}{L} \times A} \tag{3.27}$$

$$Re = \frac{\rho v L}{\mu} \tag{3.28}$$

For pipe flow, Reynold number can be written as

$$Re = \frac{\rho v D}{\mu} \tag{3.29}$$

Where, D is diameter of pipe

Froude Number (Fr):

The Froude number can be defined as the square root of the ratio of inertia force (F_i) to the gravity force (F_g). It can be derived as;

$$Fr = \sqrt{\frac{F_i}{F_g}} \tag{3.30}$$

Where, $F_i = \rho A V^2$ from Eq. (3.24)

$F_g =$ Force due to gravity $=$ m x acceleration due to gravity

$F_g = \rho \times v \times g$

$F_g = \rho \times L^3 \times g$

$$F_g = \rho \times A \times L \times g \tag{3.31}$$

Putting Eq. 3.24 and Eq. 3.31 in Eq. (3.30)

The final expression for Froude number as

$$Fr = \frac{v}{\sqrt{Lg}} \tag{3.32}$$

Euler Number (Eu):

The Euler number can be defined as the square root of the ratio of the inertia force of a flowing fluid (F_i) to the pressure force (F_P).

Mathematically, it can be written as

$$Eu = \sqrt{\frac{F_i}{F_P}} \tag{3.33}$$

$$F_i = \rho A v^2 \tag{3.34}$$

$$F_P = p \times A \tag{3.35}$$

Putting Eq. 3.34 and 3.35 in Eq. 3.33

$$Eu = \sqrt{\frac{\rho A v^2}{p \times A}}$$

$$Eu = \frac{v}{\sqrt{p/\rho}} \tag{3.36}$$

Weber Number (We):

The weber number is defined as the square root of the ratio of the inertia force (F_i) of a flowing fluid to the surface tension force (F_s). Mathematically, it can be written as;

$$We = \sqrt{\frac{F_i}{F_s}} \tag{3.37}$$

$$F_i = \text{Inertia force} = \rho A v^2 = \rho \times L^2 \times v^2 \tag{3.38}$$

$$F_s = \sigma \times L \tag{3.39}$$

Where, σ is surface tension per unit length, L is length

Putting Eq. 3.38 and 3.39 in Eq. 3.37

$$We = \sqrt{\frac{\rho \times L^2 \times v^2}{\sigma \times L}}$$

$$We = \frac{v}{\sqrt{\sigma/\rho L}} \tag{3.40}$$

Mach Number (M):

Mach number can be defined as the square root of the ratio of the inertia force (F_i) of a flowing fluid to the elastic force (F_e).

Mathematically, it is expressed as;

$$M = \sqrt{\frac{F_i}{F_e}} \tag{3.41}$$

$$F_i = \rho A v^2 \tag{3.42}$$

F_e =Elastic stress × Area

$$F_e = K \times A = K \times L^2 \tag{3.43}$$

Putting Eq. 3.42 and Eq. 3.43 in Eq. 3.41

$$M = \sqrt{\frac{\rho A v^2}{K \times L^2}}$$

$$M = \frac{v}{\sqrt{K/\rho}} \tag{3.44}$$

or,

$$M = \frac{v}{C} \tag{3.45}$$

But

$$\sqrt{\frac{K}{\rho}} = C = \text{Velocity of sound in the fluid} \tag{3.46}$$

In order to give more clarity about the understanding of the applications of dimensionless parameters, solved examples 3.8 to 3.14 are given after section 3.5.

3.5 MODEL ANALYSIS

As discussed earlier, the performance of a prototype can be predicted by the performance of a model (sized based on similarity law) which can be easily determined in laboratory. As per the similarity law, prototype and its model are called similar if they are geometrically similar, kinematically similar and dynamically similar. Further, it is necessary for dynamic similarity that prototype and its model should follow geometrical similarity and kinematic similarity laws. Based on dynamic similarity law, models are designed which includes several laws such as Reynolds model law, Froude model law, Euler model law, Weber model law and Mach model law. These laws are discussed as;

Reynolds Model Law:

This model law is developed based on the Reynolds number. The Reynolds number can be expressed as;

$$Re = \frac{\rho v D}{\mu} \tag{3.47}$$

As per the Reynolds model law, the Reynolds number for model and prototype should be same.

Mathematically, it can be expressed as

$$[Re]_m = [\mathrm{Re}]_a \tag{3.48}$$

This law is applicable for all the problems related to fluid flow where viscous force is dominating force such as pipe flow, sub-marines, airplanes etc. As an example, based on Reynolds model law, the incompressible viscous fluid flow through pipes of model and prototype can be written as

$$\frac{\rho_m v_m D_m}{\mu_m} = \frac{\rho_a v_a D_a}{\mu_a} \tag{3.49}$$

Froude Model Law:

This model law is developed based on the Froude number. The Froude number can be expressed as

$$Fr = \frac{v}{\sqrt{gL}} \tag{3.50}$$

As per the Froude model law, the Froude number for model and prototype should be same.

Mathematically, it can be expressed as

$$[Fr]_m = [\mathrm{Fr}]_a \tag{3.51}$$

This law is applicable for all the problems related to fluid flow where gravity force is dominating force such as free flow over spillways, through weirs and channels,

As an example, based on Froude model law, the free flow over spillway of model and prototype can be written as

$$\frac{v_m}{\sqrt{g_m L_m}} = \frac{v_a}{\sqrt{g_a L_a}} \tag{3.52}$$

For a case where model and porotype exist over the same level then gravitational force will be equal ($g_m = g_P$) and the above equation can be written as

$$\frac{v_m}{\sqrt{L_m}} = \frac{v_P}{\sqrt{L_a}} \tag{3.53}$$

or,

$$v_r = \frac{v_m}{v_a} = \sqrt{l_r} \tag{3.54}$$

Froude model law can be written in terms of scale ratios for various physical quantitates as follows:

(a) Time (T)

$$\text{Time} = \frac{\text{length}}{\text{velocity}} \tag{3.55}$$

$$T_r = \frac{T_a}{T_m} \tag{3.56}$$

or,

$$T_r = \frac{\left(\frac{L}{v}\right)_a}{\left(\frac{L}{v}\right)_m} \tag{3.57}$$

Final expression can be written as

$$T_r = \sqrt{l_r} \tag{3.58}$$

(b) Acceleration (A)

$$\text{A} = \frac{v}{T} \tag{3.59}$$

$$a_r = \frac{a_a}{a_m} \tag{3.60}$$

$$a_r = \frac{\left(\frac{v}{T}\right)_a}{\left(\frac{v}{T}\right)_m} \tag{3.61}$$

$$a_r = \sqrt{l_r} \times \frac{1}{\sqrt{l_r}} = 1 \tag{3.62}$$

$$a_r = 1 \tag{3.63}$$

(c) Discharge (Q)

$$Q = A \times v \tag{3.64}$$

$$Q = \frac{L^3}{T} \tag{3.65}$$

$$Q_r = \frac{Q_a}{Q_m} \tag{3.66}$$

$$Q_r = l_r^3 \times \frac{1}{\sqrt{l_r}} = l_r^{2.5} \tag{3.67}$$

(d) Force (F)

$$\mathrm{F = M \times a} \tag{3.68}$$

$$\mathrm{F} = \rho L^3 \times \frac{v}{T} = \rho L^2 \tag{3.69}$$

$$F_r = \frac{F_a}{F_m} \tag{3.70}$$

$$F_r = \frac{\rho_a}{\rho_m} \times \left(\frac{L_a}{L_m}\right)^2 \times \left(\frac{v_a}{v_m}\right)^2 \tag{3.71}$$

If the fluid used in model and prototype is same, then

$$\frac{\rho_a}{\rho_m} = 1 \text{ or } \rho_a = \rho_m \tag{3.72}$$

and then

$$F_r = \left(\frac{L_a}{L_m}\right)^2 \times \left(\frac{v_a}{v_m}\right)^2 = l_r^2 \cdot l_r = l_r^3 \tag{3.73}$$

(e) Intensity of pressure (p)

$$p = \frac{\text{Force}}{\text{Area}} = \frac{F}{A} \tag{3.74}$$

$$p = \frac{\rho L^2 V^2}{L^2} = \rho v^2 \tag{3.75}$$

$$p_r = \frac{p_a}{p_m} \tag{3.76}$$

or,

$$p_r = \frac{\rho_a v_a^2}{\rho_m v_m^2} \tag{3.77}$$

If fluid is same

$$\rho_a = \rho_m$$

then

$$p_r = l_r. \tag{3.78}$$

(f) Torque (T)

Torque (T) = Force (F) × Distance (L) = $F \times L$

∴ Torque ratio,

$$T_r^* = \frac{T_a^*}{T_m^*} \tag{3.79}$$

$$T_r^* = \frac{(F \times L)_a}{(F \times L)_m} \tag{3.80}$$

$$T_r^* = F_r \times l_r \tag{3.81}$$

$$T_r^* = l_r^3 \times l_r = l_r^4. \tag{3.82}$$

(g) Power

Power = Work per unit time

$$= \frac{F \times L}{T} \tag{3.83}$$

∴ Power ratio

$$P_r = \frac{P_a}{P_m} \tag{3.84}$$

$$P_r = \frac{\dfrac{F_a \times L_a}{T_a}}{\dfrac{F_m \times L_m}{T_m}}$$

$$P_r = \frac{F_a}{F_m} \times \frac{L_a}{L_m} \times \frac{1}{\dfrac{T_a}{T_m}}$$

$$P_r = F_r \cdot l_r \cdot \frac{1}{T_r}$$

$$P_r = l_r^3 \cdot l_r \cdot \frac{1}{\sqrt{l_r}}$$

$$P_r = l^{3.5} \tag{3.85}$$

Euler Model Law:

This model law is developed based on the Euler number. The Euler number can be expressed as

$$Eu = \frac{v}{\sqrt{p/\rho}} \tag{3.86}$$

As per the Euler model law, the Euler number for model and prototype should be same.

Mathematically, it can be expressed as

$$[Eu]_m = [\text{Eu}]_a$$

This law is applicable for all the problems related to fluid flow where pressure force is dominating force such as flow through closed pipe where only turbulence is developed while other forces are considerable negligible. This law is also used where the phenomenon of cavitation takes place. As an example, based on Euler model law, the fluid flow through closed pipe of model and prototype can be written as

$$\frac{v_m}{\sqrt{p_m/\rho_m}} = \frac{v_a}{\sqrt{p_a/\rho_a}} \tag{3.87}$$

If fluid is same ($\rho_m = \rho_a$) in model and prototype

$$\frac{v_m}{\sqrt{p_m}} = \frac{v_a}{\sqrt{p_a}} \tag{3.88}$$

Weber Model Law:

This model law is developed based on the Weber number. The Weber number can be expressed as

$$We = \frac{v}{\sqrt{\sigma/\rho L}} \tag{3.89}$$

As per the Weber model law, the Weber number for model and prototype should be same.

Mathematically, it can be expressed as

$$[We]_m = [\mathrm{We}]_a \tag{3.90}$$

This law is applicable for all the problems related to fluid flow where surface tension force is dominating force such as capillary rise in narrow passages, capillary waves developed in channels and under small head flow over weirs.

As an example, based on Weber model law, the capillary rise in narrow passages of model and prototype can be written as

$$\frac{v_m}{\sqrt{\sigma_m/\rho_m L_m}} = \frac{va}{\sqrt{\sigma_a/\rho_a L_a}} \tag{3.91}$$

Mach Model Law:

This model law is developed based on the Mach number. The Mach number can be expressed as

$$M = \frac{v}{\sqrt{\frac{K}{\rho}}} \tag{3.92}$$

As per the Mach model law, the Mach number for model and prototype should be same.

Mathematically, it can be expressed as

$$[M]_m = [\mathrm{M}]_a \tag{3.93}$$

This law is applicable for all the problems related to fluid flow where forces due to elastic compression is dominating force such as flow over the body of aeroplane and projectile and water-hammer in pipes. As an example, based on Mach model law, the flow of aeroplane through air of model and prototype can be written as

$$\frac{v_m}{\sqrt{K_m/\rho_m}} = \frac{v}{\sqrt{K_a/\rho_a}} \tag{3.94}$$

EXAMPLE 3.8

Example 3.8 Considering the dynamic similarity for two pipe flows, Find the velocity of oil (Sp. gr. of oil = 0.7) flowing in a pipe of diameter 15 cm if water is flowing through another pipe of diameter 40 cm at a velocity of 3.5 m/s. Take the viscosity value for water and oil respectively as 0.01 poise and 0.025 poise.

Solution

Given data: Two pipes having different liquids,

For pipe 1: liquid is water

Diameter, d_1 = 40 cm = 0.40 m

Velocity, v_1 = 3.5 m/sec

Viscosity, μ_1 = 0.01 poise = 0.001 N-sec/m

Density, ρ_1 = 1000 kg/m^3

For pipe 2: liquid is oil

Diameter, d_2 = 15 cm = 0.15 m

Viscosity, μ_2 = 0.025 poise = 0.0025 N-sec/m

Density, ρ_2 = sp. gr. of oil × 1000 = 700 kg/m^3

Reynolds number for two pipes is same, if pipes are having dynamically similar. Then,

$$\frac{\rho_1 v_1 d_1}{\mu_1} = \frac{\rho_2 v_2 d_2}{\mu_2}$$

Therefore, velocity of oil (v_2) is given by

$$v_2 = v_1 \cdot \frac{\rho_1}{\rho_2} \cdot \frac{d_1}{d_2} \cdot \frac{\mu_2}{\mu_1}$$

$$v_2 = 3.5 \times \frac{1000}{700} \times \frac{0.40}{0.15} \times \frac{0.025}{0.001} \times 3.5$$

V_2 = 33.33 m/s. **Ans.**

EXAMPLE 3.9

Example 3.9 Consider a drag of magnitude of 20 N experienced by a solid sphere of diameter D as 80 mm moving in water with a velocity of 5.5 m/sec. Considering the similarity law determine the velocity of sphere of diameter as 4.5 m moving in air. Find out also the drag experienced by the sphere. (ρ_{water} = 997 kg/m^3; ρ_{air} = 1.2 kg/m^3; ν_{air} = 13 ν_{water})

Solution

Given

Model data:

Diameter, D_m = 80 mm = 0.08 m,

Kinematic viscosity, ν_m = 1

Drag force, F_m = 20 N

Prototype data:

Diameter, D_p = 4. 5 m,

Kinematic viscosity, ν_p = 13

Velocity, v_p = 5.5 m/sec

Reynolds model law similarity can be used since spheres are in submerged state.

$$\left(\frac{vL}{\nu}\right)_m = \left(\frac{vL}{\nu}\right)_p$$

Velocity of sphere is given by

$$v_m = v_p \times \frac{L_p}{L_m} \times \frac{\nu_m}{\nu_p}$$

$$= 5.5 \times \frac{4.5}{0.08} \times \frac{1}{13}$$

$$= 23.79 \text{ m/s}$$

Now, ratio of the drag for the model and the prototype is given by

$$\frac{F_m}{F_p} = \frac{\rho_m}{\rho_p} \times \left(\frac{L_m}{L_p}\right)^2 \times \left(\frac{v_m}{v_p}\right)^2$$

$$= \frac{1000}{1.2} \times \left(\frac{100}{4.5 \times 1000}\right)^2 \times \left(\frac{5}{23.79}\right)^2$$

$$\frac{F_m}{F_p} = 0.0181$$

Drag force experienced by the sphere is given by

$F_p = \frac{20}{0.0181} = 1100$ N. **Ans.**

EXAMPLE 3.9

Example 3.10 A model of a spillway was tested for a discharge of 1.8 m^3/s with a velocity of 1.25 m/sec. Applying similarity, find out velocity and discharge for a prototype spillway developed taking the size ratio as 35:1.

Solution.

Given

Model data:

Discharge, Q_m = 1.8 m^3/sec

Velocity, v_m = 1.25 m/sec

EXAMPLE 3.10

EXAMPLE 3.10

Size ratio, l_r = 35:1

Froude model law can be used for dynamic similarity.

$$\frac{v_a}{v_m} = \sqrt{l_r} = \sqrt{35}$$

$$= 5.916$$

Velocity over prototype, v_a

$$v_a = v_m \times 5.916 = 1.25 \times 5.916$$

$$= 7.39 \text{ m/s } \textbf{Ans.}$$

Now, for discharge over prototype (Q_P)

$$\frac{Q_a}{Q_m} = l_r^{2.5} = (35)^{2.5}$$

$$Q_a = 1.8 \times 35^{2.5}$$

$$= 13044.95 \text{ m}^3\text{/sec } \textbf{Ans.}$$

EXAMPLE 3.11

Example 3.11 The scale of a spillway model is considered as 1:10 of its prototype. Considering viscous force and surface tension as negligible, find out the flow rate through the model if the flow rate through the prototype is 1100 m^3/s. Determine the energy dissipation in the prototype if the energy dissipation in model occurs as 300 watts.

Solution

Given

Size ratio, l_r = 1:10

Flow rate through prototype, Q_a =1100 m^3/s

Energy dissipation in model, P_m = 300 W

Since gravity is the principal force, Froude model law can be used.

$$Q_r = l_r^{5/2} g_r^{1/2}$$

$$Q_r = \frac{Q_m}{Q_a} = \left(\frac{1}{10}\right)^{5/2} \times 1 \quad (g_r = 1)$$

Flow rate through model, Q_m

$$Q_m = 1100 \times \left(\frac{1}{10}\right)^{5/2}$$

$$= 3.478 \text{ m}^3/\text{s}$$

Now, based on Froude model law, scale ratio for power is given by

$$P_r = \rho_r l_r^{7/2} g_r^{3/2}$$

$$\frac{P_m}{P_a} = 1 \times \left(\frac{1}{10}\right)^{7/2} \times 1 \quad (\rho_r = 1)$$

Energy dissipation in prototype (P_a) is given by

$$P_a = 300 \times (10)^{7/2}$$

= 948.68 kW. **Ans.**

EXAMPLE 3.11

Example 3.12 The parameters for a prototype and model of over flow structure are given as: the length of prototype is 600 m, discharge through prototype is 4500 m^3/s, channel length in the laboratory is 0.6 m wide and model to prototype size ratio as 1:18. Find out the flow rate through the model, neglecting the effect of viscosity and surface tension.

Solution

Given

Model data:

Length, $L_m = 0.6$ m

Prototype data:

Length, $L_a = 600$ m

Discharge, $Q_a = 4500$ m^3/s

Size ratio, $l_r = 1:18$

Based on Froude model law,

$$v_r = (l_r)^{1/2}$$

$$v_r = \left(\frac{1}{18}\right)^{1/2}$$

Now, discharge is given by

$$Q_r = A_r v_r$$

$$\frac{Q_m}{Q_a} = (v_r l_r) \times v_r$$

$$Q_m = 4500 \times \left(\frac{0.6}{600} \times \frac{1}{18}\right) \times \left(\frac{1}{18}\right)^{1/2} \quad \left(l_r = \frac{L_m}{L_a}\right).$$

$Q_m = 0.0589$ m^3/s. **Ans.**

EXAMPLE 3.12

EXAMPLE 3.13

Example 3.13 During testing in water of a model of an aero plane having the scale size of 1:40 the pressure drop was found as 75 N/ cm^2. Determine the pressure drop of the prototype in air (ρ_{air}= 1.2 kg/m^3, μ_{water} = 0.01 poise and μ_{air} = 0.00018 poise)

Solution.

Given

Model data (water):

Pressure, $p_m = 75\ \text{N/cm}^2 = 75 \times 10^4\ \text{N/m}^2$

Viscosity, μ_m = 0.01 poise = 0.001 Ns/m

Density, $\rho_m = 1000\ \text{kg/m}^3$

Prototype data (air):

Viscosity, μ_a = 0.00018 poise = 0.000018 Ns/m

Density, $\rho_a = 1.2\ \text{kg/m}^3$

Scale size, $l_r = 40$

Based on Reynold's number model law, we get

$$\frac{\rho_m v_m L_m}{\mu_m} = \frac{\rho_a v_a L_a}{\mu_a}$$

or,

$$\frac{v_m}{v_a} = \frac{\rho_a}{\rho_m} \times \frac{L_a}{L_m} \times \frac{\mu_m}{\mu_a}$$

$$\frac{v_m}{v_a} = \frac{1.2}{1000} \times \frac{40}{1} \times \frac{0.0001}{0.000018}$$

$$\frac{v_m}{v_P} = 2.667$$

Now, based on Euler number model law equal, we get

$$\frac{v_m}{\sqrt{\frac{p_m}{\rho_m}}} = \frac{v_a}{\sqrt{\frac{p_a}{\rho_a}}}$$

$$\frac{v_m}{v_a} = \sqrt{\frac{p_m}{p_a}} \times \sqrt{\frac{\rho_a}{\rho_m}}$$

$$2.667 = \sqrt{\frac{75\times10^4}{p_a}} \times \sqrt{\frac{1.2}{1000}}$$

pressure drop of the prototype in air is given by

$p_a = 0.01265\ \text{N/cm}^2$. **Ans.**

EXAMPLE 3.13

Example 3.14 What will be the velocity and discharge through a prototype of a spillway for the following data: (i) model – prototype scale as 1:35 (ii) velocity over model 1.8 m/sec (iii) discharge through model 2.35 m^3/s

Solution

Given:

Model data:

Velocity, v_m = 1.8 m/sec

Discharge, Q_m = 2.35 m^3/s

Scale ratio, l_r = 1:35

Velocity ratio is given by

$$\frac{v_a}{v_m} = \sqrt{l_r} = \sqrt{35}$$

$$v_a = v_m \times \sqrt{35}$$

$$= 1.8 \times \sqrt{35}$$

v_a = 10.648 m/sec. **Ans.**

Now, for discharge ratio is given by

$$\frac{Q_a}{Q_m} = l_r^{2.5}$$

$$Q_a = Q_m \times L_r^{2.5}$$

$$Q_a = 2.35 \times 35^{2.5}$$

Q_a= 17030.9 m^3/s. **Ans.**

UNIT SUMMARY

- Dimensional analysis deals with the relationship between the physical quantities which are analysed with the help of dimensions and units of measurements.
- In several fluid mechanics problems, the relationship between dimension variables of physical system can be established by two methods:

  Rayleigh's method

  Buckingham's π-theorem method

Mathematically, Rayleigh's method can be expressed as follow

$$X = CX_1^a \cdot X_2^b \cdot {X_3}^c$$

Where, C is constant and a, b and c are arbitrary powers.

Buckingham π-theorem minimize the drawback faced in Rayleigh's Method having limitation of four number of variables.

- The model and prototype can be said similar if these are: geometrically similar, kinematically similar and dynamically similar.
- Geometric similarity

  In case of geometric similarity, the ratio of two physical dimensions of an actual (prototype) is equal to the ratio of corresponding dimensions of model.

  Length scale ratio:

$$l_r = \frac{L_m}{L_a} = \frac{B_m}{B_a} = \frac{D_m}{D_a}$$

  Area scale ratio:

$$A_r = l_r^2$$

  Volume scale ratio:

$$V_r = l_r^3$$

- Kinematic similarity

  It says that the ratio of two velocities of fluid flow through an actual (prototype) system is equal to the ratio of corresponding velocity through model.

  Velocity scale ratio:

$$v_r = l_r/T_r$$

  Acceleration scale ratio:

$$a_r = \frac{a_m}{a_a} = \frac{l_r}{T_r^2}$$

  Discharge scale ratio:

$$Q_r = \frac{Q_m}{Q_a} = \frac{l_{r^3}}{T_r}$$

- Dynamic similarity

  If the force ratios acting at homologous points in the model and prototype are equal, there is dynamic similarity between the model and prototype.

  (i) Reynolds number

$$Re = \frac{\rho v D}{\mu}$$

  (ii) Froude number

$$Fr = \frac{v}{\sqrt{Lg}}$$

  (iii) Euler number

$$Eu = \frac{v}{\sqrt{p/\rho}}$$

(iv) Mach number

$$M = \frac{v}{\sqrt{K/\rho}}$$

(v) Weber number

$$\text{We} = \frac{v}{\sqrt{\sigma/\rho l}}$$

- A variety of model laws have been developed depending on the significance of each force of the different phenomena, which are listed below

(i) Reynolds Model Law

$$\frac{\rho_m v_m l_m}{\mu_m} = \frac{\rho_a v_a l_a}{\mu_a}$$

(ii) Froude Model Law

$$\frac{v_m}{\sqrt{g_m l_m}} = \frac{v_a}{\sqrt{g_a l_a}}$$

(iii) Euler Model Law

$$\frac{v_m}{\sqrt{p_m/\rho_m}} = \frac{v_a}{\sqrt{p_a/\rho_a}}$$

(iv) Mach Model Law

$$\frac{v_m}{\sqrt{K_m/\rho_m}} = \frac{v_a}{\sqrt{K_a/\rho_a}}$$

(v) Weber Model Law

$$\frac{v_m}{\sqrt{\sigma_m/\rho_m l_m}} = \frac{v_a}{\sqrt{\sigma_a/\rho_a L_a}}$$

EXERCISES

Multiple Choice Questions

3.1 Equations that are dimensionally homogeneous are applicable for

(a) C.G.S. system only

(b) M.K.S. and SI systems

(c) FPS. system only

(d) all systems of units

3.2 Which of following is primary quantity?

(a) Volume (b) Speed (c) Density (d) Mass

3.3 Dimensions whose unit is independent of any other dimension are called ________

(a) Absolute dimension (b) Independent dimension

(c) Dependent dimension (d) Fundamental dimension

3.4 Which of the following methods is used to determine a functional relationship with respect to a parameter?

(a) Doppler effect (b) Newton's laws

(c) Rutherford's method (d) Rayleigh's method

3.5 Rayleigh's method should be presented in following formats?

(a) $D = f(lpv)$ (b) $D = f$ (c) $D = (l, \rho, \mu, v, g)$ (d) $D = f(l, \rho, \mu, v, g)$

3.6 Which of the following is similar to Rayleigh's method?

(a) Ionization method (b) Conveyor method

(c) Dead weight method (d) Buckingham method

3.7 What is the number of π-terms based on six physical quantities and three fundamental unit is

(a) 4 (b) 3 (c) 2 (d) 1

3.8 Among the following, which is not considered a core group?

(a) Flow characteristics (b) Fluid property

(c) Geometric property (d) Time

3.9 A model analysis has the following advantages:

(a) It is possible to predict alternative designs

(b) It is not possible to determine the relationships between the variables

(c) Shear stress to thermal diffusivity

(d) It is impossible to predict performance

3.10 The concept of similitude can be applied to the testing of ________

(a) Chemical models (b) Engineering models

(c) Physical models (d) Mathematical models

3.11Dynamic similarity between prototype and model is the

(a) similarity of forces (b) similarity of lengths

(c) similarity of motion (d) None of the above

3.12 There is a similarity between the motion of the prototype and the motion of the model

(a) Design similarity (b) Kinematic similarity

(c) Potential similarity (d) Dynamic similarity

3.13 Kinematic similarity between prototype and model is the

(a) similarity of streamline pattern (b) similarity of shape

(c) similarity of discharge (d) none of the above

3.14 The following does not constitute a criterion of similitude?

(a) Conditional similarity (b) Dynamic similarity

(c) Kinematic similarity (d) Geometric similarity

3.15 Which of the following is considered to be a standard scale for a similitude?

(a) 1:100 (b) 1:25 (c) 1:50 (d) 1:250

3.16 Which of following does belong to dimensionless number?

(a) Cartesian (b) Mach (c) Froude (d) Reynolds

3.17 The formula for elastic force is as follows:

(a) Elastics stress × Elastic (b) Elastic stress × area

(c) Elastic strain/area (d) Elastic stress/area strain

3.18 ________ refers to ratio of inertia to viscous force.

(a) Mach number (b) Reynolds number

(c) Weber number (d) Froude number.

3.19 Which of the following term represents the square root of ratio between inertia and pressure force?

(a) Froude number (b) Euler number

(c) Mach number (d) Reynolds number.

3.20 Mach number is equal to the square root of ratio between the

(a) inertia to elastic force (b) inertia to surface tension force

(c) inertia to pressure force (d) none of the above.

Answers of Multiple-Choice Questions

3.1 (d), 3.2 (d), 3.3 (d), 3.4 (d), 3.5 (d), 3.6 (d), 3.7 (b), 3.8 (b), 3.9 (a), 3.10 (b), 3.11 (a),3.12 (b), 3.13 (a), 3.14 (a), 3.15 (b), 3.16 (a), 3.17 (b), 3.18 (b), 3.19 (b), 3.20 (a).

Short and Long Answer Type Questions

3.1 Explain the meaning of the terms in dimensional analysis and model analysis.

3.2 Describe the concept of dimensionally homogeneous equations

3.3 Describe the three primary objectives of dimensional analysis.

3.4 Give the dimensions of: (i) Force (ii) Viscosity (iii) Power and (iv) Kinematic viscosity.

3.5 In order to demonstrate your work, list the fundamental dimensions of each of the following thermodynamics variables: (a) Energy, E (b) Specific Energy, e = E/m (c) Power, W.

3.6 To establish the functional relationship between the various parameters, what are the various methods for the dimensional analysis can be used, that affect the physical phenomenon.

3.7 Explain the Rayleigh's method used for dimensional analysis.

3.8 How do you define the repeating variables? How is this selection made using dimensional analysis?

3.9 Define the various applications of model testing?

3.10 Describe the various advantages of model testing?

3.11 How do you define the geometric, kinematic, and dynamic similarities? Is it possible to achieve these similarities? If not, why?

3.12 Define the dimensionless numbers?

3.13 Show that ratio of inertia force to viscous force gives the Reynolds number.

3.14 Explain the following dimensionless numbers and state their implications for fluid flow issues. (i) Reynolds number, (ii) Froude number, and (iii) Mach number,

3.15 Define the terms: model, prototype, model analysis, hydraulic similitude.

Numerical Problems

3.16 Show that if capillary rise h depends on the fluid surface tension σ, specific weight ω, and tube radius r.

$$\frac{h}{r} = \emptyset\left(\frac{\sigma}{\omega r^2}\right)$$

3.17 Show that the equations can be used to express the shear stress (τ) in a fluid flowing through a pipe:

$$\tau = \rho V^2 \emptyset\left[\frac{\mu}{\rho D v}, \frac{k}{D}\right]$$

where, D as diameter of pipe, ρ as mass density, V as velocity, μ as viscosity, and k as height of roughness projection.

3.18 Using dimensional analysis find an expression for the force having exerted by a flowing fluid on a stationary body depends upon velocity (v) of the fluid, viscosity (μ) of the fluid, length (L) of the body, density (ρ) of fluid, and acceleration (g) due to gravity.

3.19 A porotypes pipe with diameter 2 m is required to transport an oil of sp. gr. 0.8 and viscosity 0.04 poise at the rate of 4 m^3/s. Tests were conducted on a 20 cm diameter pipe using water at 20°C. Find the velocity and rate of flow in the model. Viscosity of water at 20°C = .01 poise.

3.20 What is the force ratio, F, for the different similarity criteria (a) Froude number, (b) Reynolds number, (c) Weber number, and (d) Mach number, assuming that the same fluid at the same temperature is to be used in the model as in the prototype?

3.21 A spillway 8.2 m high and 160 m long discharges 2250 m^3/s under a head of 5 m. If a 1:16 model of the spillway is to be constructed, find the model dimensions, head over the model and the model discharge.

3.22 A flying boat model in scale 1:20 is pulled through the water. The prototype is moving at a speed of 16 m/s in seawater with a density of 1024 kg/m^3. Find the model's speed. If the prototype's wave resistance is 500 N, then calculate the model's wave resistance.

3.23 Show that the ratio of discharges per unit width of the spillway is given by $\left(\frac{1}{75}\right)^{3/2}$ if the model prototype ratio is 1:75.

3.24 An air duct is to be modelled to a scale ratio 1:20 and tested with water which is 800 times denser and 50 times viscous than air. When tested under dynamically similar conditions, the pressure drop between two sections in the model is 240 kPa. What is the corresponding pressure drop in the prototype?

3.25 The velocity and discharge in a spillway's 1: 30 model are 1.5 m/sec and 2.0 m3/sec, respectively. In the prototype, determine the corresponding velocity and discharge

REFERENCES AND SUGGESTED READINGS

List of some of the books is given below which may be used for further learning of the subject:

1. E. Mosonyi, Water Power Development, Vol. I and II, Nem Chand and Brothers, 2009.
2. J. Lal, Hydraulic Machines, 3rd edition (reprint), Metropolitan Book Co. Private Limited, 2002.
3. G. Brown, Hydro-electric Engineering Practise, Vol. II, CBS Publication, 1984.
4. Yunus A. Çengel, Fluid Mechanics: Fundamentals and Applications, McGRAW-HILL publication, 2006.
5. P. N. Modi, Hydraulics & Fluid Mechanics including Hydraulics Machines, Rajsons Publications Pvt. Ltd., 2014.
6. White, Frank M, Fluid Mechanics, 7th edition, McGraw-Hill 2009.
7. Philip J. Pritchard, Fox and McDonald's Introduction to Fluid Mechanics, 8th edition, John Wiley & Sons 2011.
8. Robert W. Fox, Alan T. McDonald and Philip J. Pritchard, Introduction to fluid mechanics 6th edition 2004.

4 Fluid Machines – Pumps

UNIT SPECIFICS

Through this unit we have discussed the following aspects:

- *Definition of pumps and their classifications;*
- *Euler's equation and theory of rotodynamic machines;*
- *Different efficiencies and velocity components of pumps;*
- *Working principle of centrifugal pumps and work done by the Impeller;*
- *Performance curves and cavitation in pumps;*
- *Selection of pumps and concept of pump as turbine;*
- *Working principle of positive displacement pump;*

The practical applications of the topics are discussed for generating further curiosity and creativity as well as improving problem solving capacity.

Besides giving a large number of multiple-choice questions as well as questions of short and long answer types marked in two categories following lower and higher order of Bloom's taxonomy, assignments through a number of numerical problems, a list of references and suggested readings are given in the unit so that one can go through them for practice. It is important to note that for getting more information on various topics of interest some QR codes have been provided in different sections which can be scanned for relevant supportive knowledge.

RATIONALE

This fundamental unit on pumps helps students to get a primary idea about the working principle, characteristics and selections of pumps for field application of these under different operating conditions. It explains the Euler's equation to know the fluid motion through centrifugal pumps. It then explains the various velocity components at the entry and exit of the rotor (impeller) to understand the procedure in order to determine the work done by the impeller. To understand the behaviour of pumps under given conditions, performance curves are discussed. Concept of pump used as turbine is also included in the unit and stepwise selection of pumps for different site conditions is given.

Pumps is an important unit under fluid machines essentially deals transfer of fluids from one location to another location by receiving the mechanical energy. This permits to understand the working principle, classification, characteristics and selection of pumps to help the students in studying these machines under advanced courses and with industries.

PRE-REQUISITES

Science: Properties of fluids especially water such as density, specific gravity, viscosity, flow velocity etc. (Class XII)

Physics: Fundamental of energy conversion process such as conservation of mass (continuity equation) and energy (Bernoulli's equation) (Class XII)

Mathematics: Vector analysis such as velocity diagrams and various forces (Class XII)

UNIT OUTCOMES

List of outcomes of this unit is as follows:

U4-O1: Describe the working principle and classification of centrifugal pumps

U4-O2: Draw the velocity triangles and evaluate the work done by the Impeller

U4-O3: Define the various efficiencies and cavitation in pumps

U4-O4: Describe the performance curves of pumps, selection procedure of pumps and concept of pump as turbine

U4-O5: Describe the working principle of positive displacement pump and solve pumps related problems

Unit-4 Outcomes	EXPECTED MAPPING WITH COURSE OUTCOMES (1- Weak Correlation; 2- Medium correlation; 3- Strong Correlation)					
	CO-1	CO-2	CO-3	CO-4	CO-5	CO-6
U4-O1	3	2	3	1	-	2
U4-O2	3	1	2	2	3	1
U4-O3	2	3	-	3	2	1
U4-O4	-	2	3	1	3	2
U4-O5	1	3	3	-	2	1

4.1 ROTODYNAMIC MACHINES

In practical applications, we need to develop the machines which work on the mechanism for converting mechanical power into pressure energy & kinetic energy to a fluid (liquid or gas) in case of pumps, while, pressure(potential) energy & kinetic energy of fluid into mechanical power in case of turbines. These machines are known as rotodynamic machines which use either liquid or gas as the working fluid. Under this section, a brief about the classification and theory of rotodynamic machines deals with liquid as fluid are discussed.

4.1.1 Theory of Rotodynamic Machines

The theory of rotodynamic machine can be discussed based on the "Newton's second law of motion". In rotodynamic machines the most important component is the rotor (impeller or runner), whose motion with respect to fluid flow during operation of the machine needs to be understood. Due to change in velocity, there will be change in momentum, accordingly, a force in the direction tangential to the rotor will be acting. In case of a turbine, the tangential force of the fluid is responsible to do the work by the moving rotor. However, the tangential force is increased in case of pumps and compressors, where the pressure head is created by transferring the energy. In order to discuss the theory of rotodynamic machines, different velocity components (referring different locations over the rotor as '1' and '2) observed during the interaction between fluid and the rotor are shown in Fig. 4.1.

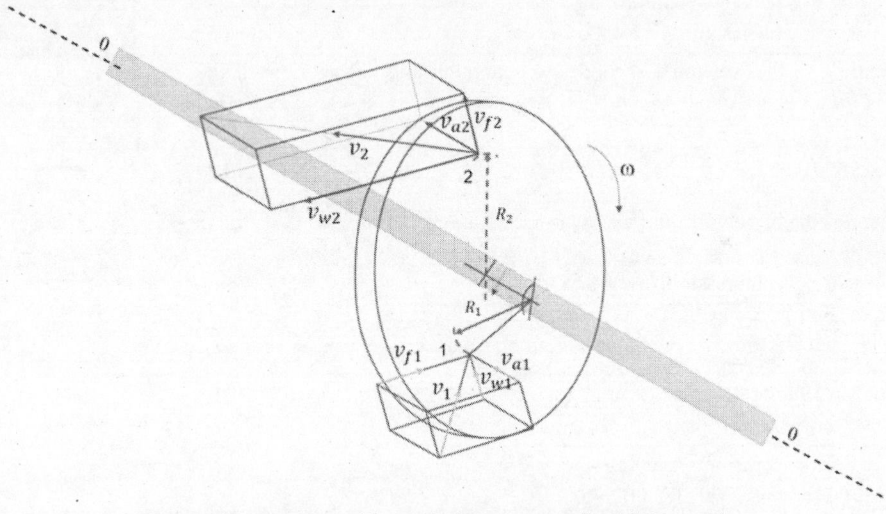

Fig. 4.1: Components of flow velocity in a generalised fluid machine.

Considering, R_1 and R_2 are the radii of the rotor corresponding to 1 (entry) and 2 (exit) and corresponding to these points about the axis 0-0, ω is angular velocity of the fluid. The assumptions made to analyse the energy transfer are; (i) flow is in steady state, (ii) constant rate of heat and work transfer, and (iii) velocity is uniform over the considered mass flow of a finite area.

The velocity at any point is resolved into three mutually perpendicular components as; axial (v_a), radial (v_f) and tangential (v_w). Due to change in v_a component magnitude, there will be change in rotor momentum in axial direction, which will result in axial force in terms of axial thrust on stationary rotor casing. On the other hand, v_f component change in magnitude which causes a change in momentum in radial direction.

In case of pumps and compressors there is an increase in the tangential momentum of the fluid and therefore work is absorbed by the fluid from moving rotor unlike in case of turbines, where work is done by the fluid to the moving rotor. In this case the power transfer to or from the fluids due to the change in the angular momentum of the fluids by rotating component of the machines. This phenomenon can be discussed based on Euler's equation of motion.

4.1.1.1 Euler's equation of motion

In fluid dynamics, the Euler's equation governs the motion of a compressible, inviscid fluid. When a fluid is in motion, various forces as listed below will be acting on it.

(i) Pressure force, F_p
(ii) Gravity force, F_g
(iii) Friction force due to viscosity, F_v
(iv) Force due to turbulence force, F_t
(v) Force due to compressibility, F_c
(vi) Force due to elasticity, F_e

SCAN ME

For better understanding euler's equation

According to the Newton's second law of motion and considering all the above forces in a given direction 's', the equation of motion can be expressed as

$$F_{ps} + F_{gs} + F_{vs} + F_{ts} + F_{cs} + F_{es} = \rho\ dA\ ds\ a_s \tag{4.1}$$

Where, ρ is the density of fluid and a_s is acceleration.

Generally, forces due to viscosity, turbulence, compressibility and elasticity are very small for viscous fluid having low velocity and will be neglected to express the Euler's equation as follows

$$F_{ps} + F_{gs} = \rho\ dA\ ds\ a_s \tag{4.2}$$

Hence, Euler's equation represents the relationship among velocity, pressure, and density of a steady state fluid (ideal) flow along a streamline.

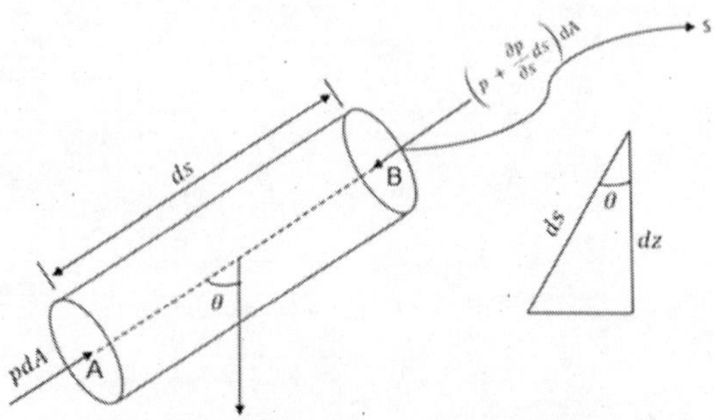

Fig. 4.2: Cylindrical element of fluid flow

In order to drive the equation in generalised form, let the fluid is considered in flowing stage from point A to point B in a stream line as shown in Fig. 4.2. Across the cylindrical cross section of fluid, considering a small section of cross-sectional area dA of length ds, the pressure force in the direction of flow will be as pdA.

Therefore, the forces acting on the cylindrical element of fluid flow will be as; pressure force $\left(p + \frac{\partial p}{\partial s} ds\right)$ dA in the flow direction while in the opposite direction of fluid flow, weight of fluid element ($\rho g dA dS$) will be there.

If θ is the angle between the direction of fluid flow and the line of action of weight of the fluid element, the element of fluid flow will experience the net force in the considered direction S which can be expressed as

$$p\,dA - \left(p + \frac{\partial p}{\partial s} ds\right) dA - \rho g dA \cos\theta = \rho\, dA\, ds\, a_s \tag{4.3}$$

Where, a_s is acceleration in S direction

$$a_s = \frac{dv}{dt} = \frac{\partial v}{\partial s}\frac{ds}{dt} + v\frac{\partial v}{\partial s} + \frac{\partial v}{\partial t} \tag{4.4}$$

Where, v is function of s and t.

If flow is in steady state, then

$$\frac{\partial v}{\partial t} = 0 \tag{4.5}$$

So, Eq. (4.4) becomes

$$a_s = v\frac{\partial v}{\partial s} \tag{4.6}$$

$$\cos\theta = \frac{dz}{ds} \tag{4.7}$$

The above expressions can be written in simplified form and expressed by Eq. 4.8 which is known as Euler's equation of motion

$$\partial p/\rho + v dv + g dz = 0 \tag{4.8}$$

4.1.2 Pumps

Pump can be defined as a prime mover used to convert mechanical energy obtained from driver or electrical motor into pressure head or kinetic energy or both to the fluids. Pumps are generally used for raising liquids from a low pressure at inlet to high pressure at outlet of the pump. Based on the working principle or the manner in which the mechanical energy is converted to the fluid, pumps are categorized as; (i) non-positive displacement or rotodynamic pumps, and (ii) positive displacement pumps. The classifications of pumps are shown in Fig. 4.3.

4.1.2.1 Rotodynamic pumps

In case of non-displacement or rotodynamic pumps, the energy of fluids is increased due to kinetic energy, pressure energy and centrifugal action. Primarily, these pumps transfer mechanical energy obtained from the driver to the fluid by increasing the fluid kinetic energy and then convert this into pressure head of fluid flowing through the pump. It is therefore, the rotodynamic pumps can also be defined as a fluid machine where kinetic energy is converted into pressure head by means of a rotating impeller, propeller or rotor. In general, the rotodynamic pumps are classified as; (i) radial flow, (ii) mixed flow, and (iii) axial flow pumps. The flow direction over the impeller in different types of pumps is shown in Fig. 4.4.

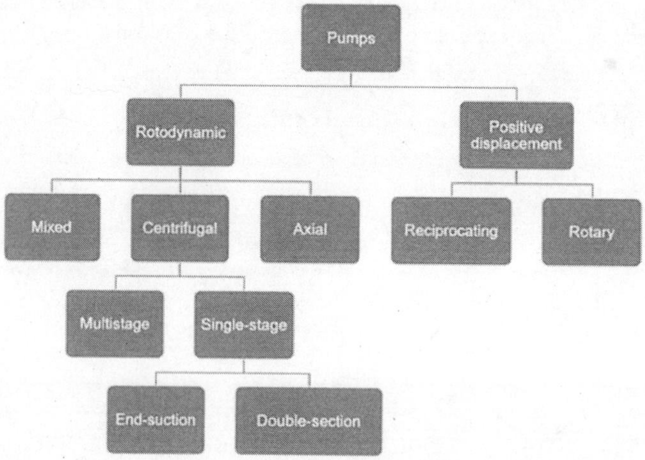

Fig. 4.3: Classification of pumps

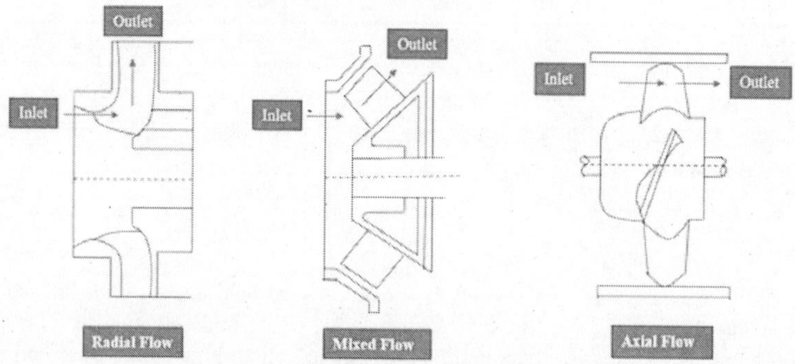

Fig. 4.4: Radial, mixed and axial flow impellers

4.1.2.2 Centrifugal pumps

In practice, among all the rotodynamic pumps, centrifugal pumps are used mostly to transport liquid. Centrifugal pumps can be defined as the fluid machine which is used to convert mechanical power of driver into pressure head in the liquid. This pressure head is used to transport liquids from one location to another location under the required pressure considering all losses in liquid conducting system. In case of centrifugal pumps, the liquid enters at the entry along the shaft (parallel direction) to the impeller and leaves in radial direction with a high pressure, as shown in Fig. 4.5. A centrifugal pump basically consists of an impeller along with shaft & bearings and a volute casing. Impeller is considered as most important component which consists of vanes between two discs.

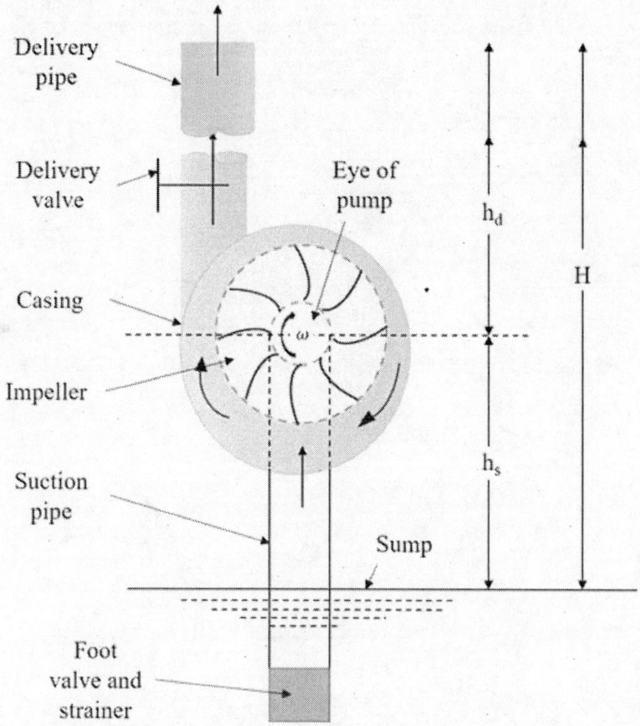

Fig. 4.5: Schematic diagram of a centrifugal pump

Out of these two discs, one disk is open in the centre (known as 'eye' of the pump) through which the liquid is suck in over the impeller vanes. These rotating vanes create the centrifugal force in the liquid and leave it across the periphery of the impeller at exit of the casing. The exit which is known as the delivery of the pump is connected to the delivery pipe to transport the liquid under pressure at the desired location. Based on the types of impellers, centrifugal pumps are classified into three types (Fig. 4.6) as; (i) backward-inclined blades, which are most efficient and commonly used, (ii) straight or radial blades, are the simplest and produce the largest pressure rise for a wide range of volume flow rates, and (iii) forward-inclined blades, generally handle lower volume with higher efficiency.

4.2 WORKING PRINCIPLE AND OPERATION OF CENTRIFUGAL PUMP

The working principle of a centrifugal pump is similar of other rotodynamic machines where pressure is created to lift the liquid by rotating impeller in the casing. The creation of pressure can be possible only if the impeller inside the casing is completely submerged in liquid. Otherwise, the presence of an air pocket inside the pump may keep the liquid away from the contact of the impeller. The pressure developed by the impeller is directly proportional to the density of fluid. If air is in contact with impeller, the pressure developed will be very small. This process is known as the 'Priming' of the pump. After completing the priming, the driver or the electric motor is started to rotate the impeller of the pump.

Initially, the delivery valve is kept closed and once the required speed of the pump is achieved, the valve is opened for the required discharge and head. Fig. 4.7 shows the cross-sectional view of volute casing of a centrifugal pump.

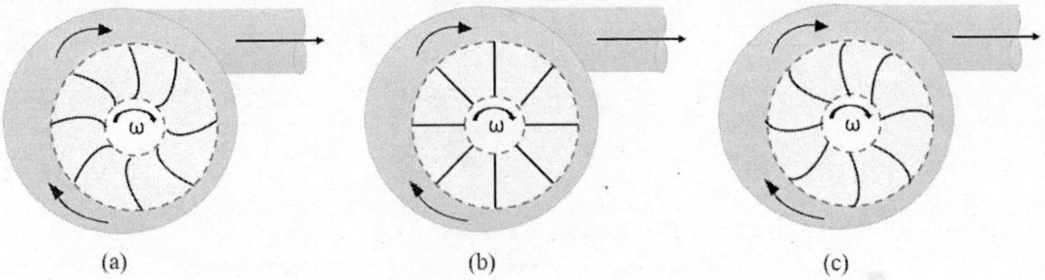

Fig. 4.6: Types of centrifugal pump (a) with backward-inclined blades (b) with radial blades (c) with forward-inclined blades

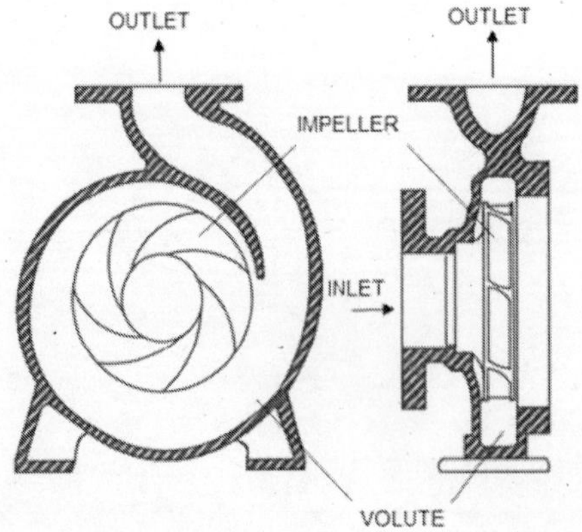

Fig. 4.7: Cross- sectional view of volute casing and impeller of a centrifugal pump

4.2.1 Work Done by the Impeller on Liquid

In order to determine the work done by a centrifugal pump, the components of different velocities need to be determined. The resultant of these velocities can be obtained by drawing the velocity diagrams or velocity triangles. The work done by the impeller can be discussed by an expression which can be derived by drawing the velocity triangles considering different velocities. Followings velocities are considered to derive the expression for work done.

(i) u_1 and u_2 are the peripheral velocity of impeller's vane tip angle at inlet and outlet, respectively.
(ii) v_1 and v_2 are the absolute velocity of water at inlet and outlet, respectively.
(iii) w_1 and w_2 are the relative velocity of water at inlet and outlet, respectively.
(iv) v_{u_1} and v_{u_2} are the whirl velocities at inlet and outlet, respectively.
(v) v_{f_1} and v_{f_2} are the flow velocities at inlet and outlet, respectively.

(vi) α_1 and α_2 are the angles made by absolute velocity at inlet and outlet with the direction of motion of vane, respectively.

(vii) β_1 and β_2 are the angles made by relative velocity at inlet and outlet with the direction of motion of vane, respectively.

Fig. 4.8. shows the velocity triangles at the inlet and outlet tips of the vanes fixed to an impeller.

As discussed earlier, the water enters the impeller radially at inlet, accordingly the absolute velocity of water at inlet with an angle, α of 90° in the direction of motion of the impeller at inlet. Considering, α as 90° and $V_{w1} as$ 0, the velocity triangles are drawn as:

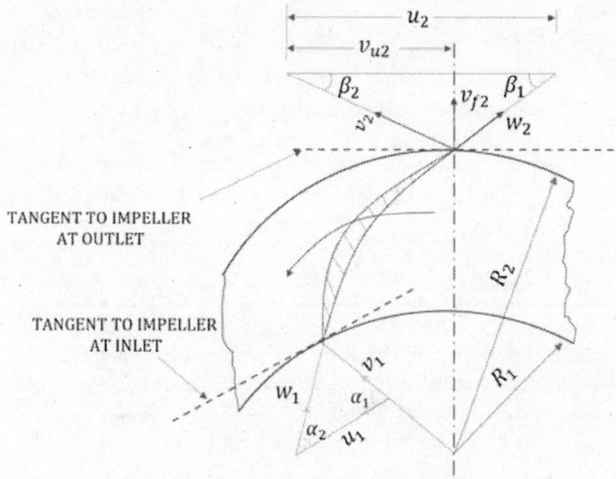

Fig. 4.8: Velocity triangle at the inlet and outlet

Let, N = Speed of the impeller in r.p.m.,

D_1 = Diameter of impeller at inlet

$$u_1 = \frac{\pi D_1 N}{60} \tag{4.9}$$

D_2 = Diameter of impeller at outlet,

u_2 = Tangential velocity of impeller at outlet

$$u_2 = \frac{\pi D_2 N}{60} \tag{4.10}$$

Work done by the impeller on the water per second per unit weight of water striking per second is given by

$$= \frac{1}{g}\left(v_{u_1} u_2 - v_{u_2} u_1\right) \tag{4.11}$$

For the conditions, when the water enters with an absolute velocity, $v_{u_1} = 0$ in the radial direction then $\alpha = 90^o$.

$$= \frac{1}{g} v_{u_2} u_2 \tag{4.10}$$

Work done by impeller on water per second is given by

$$= \frac{W}{g} v_{u_2} u_2 \tag{4.11}$$

Where,

W = Weight of water = $\rho \times g \times Q$

Q = Volume of water

$$Q = \text{Area} \times \text{Velocity of flow} = \pi D_1 B_1 \times v_{f_1} \tag{4.12}$$

$$Q = \pi D_2 B_2 \times v_{f_2} \tag{4.13}$$

In the above expression, B_1 and B_2 are width of impeller at inlet and outlet, respectively. While, v_{f_1} and v_{f_2} are the corresponding flow velocities.

EXAMPLE 4.1

Example 4.1 Find out the work done by the impeller with the help of velocity triangles for unit discharge, if the water enters the impeller radially with constant flow velocity. Consider impeller internal and external diameters as 150mm and 350mm respectively with vane tip angles at inlet and outlet respectively as 25° and 35°. Assume, the pump speed as 1500 r.p.m.

Solution

Given data,

D_1 = 150 mm = 0.15 m

D_2 = 350 mm = 0.35 m

N = 1500 r.p.m.

$\alpha_2 = 25°$

$\alpha_1 = 35°$

Water enters radially means,

$\alpha_1 = 90°$ and $v_{u1} = 0$

$v_{f1} = v_{f2}$

Tangential velocity of impeller at inlet and outlet are,

$$u_1 = \frac{\pi D_1 N}{60} = \frac{\pi \times 0.15 \times 1500}{60} = 11.78 \text{ m/s}$$

$$u_2 = \frac{\pi D_2 N}{60} = \frac{\pi \times 0.35 \times 1500}{60} = 27.49 \text{ m/s}$$

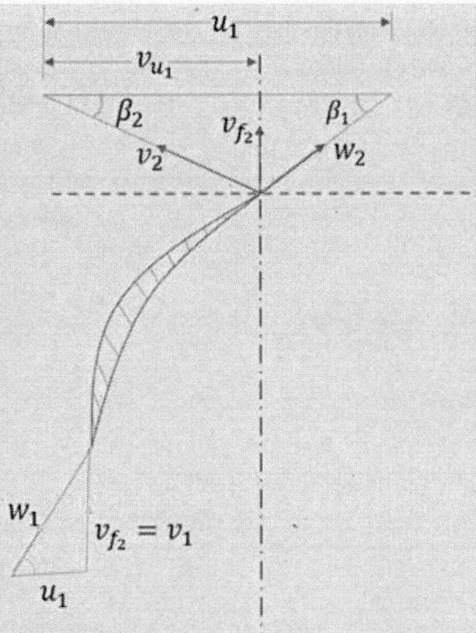

From inlet velocity triangle,

$$\tan(\alpha_2) = \frac{v_{f1}}{u_1} = \frac{v_{f1}}{11.78}$$

$$v_{f1} = 11.78\tan(\alpha_2) = 11.78 \times \tan(25) = 5.493 \text{ m/s}$$

$$v_{f2} = v_{f1} = 5.493 \text{ m/s}$$

From outlet velocity triangle,

$$\tan(\beta_1) = \frac{v_{f2}}{u_2 - v_{u2}} = \frac{5.493}{27.49 - v_{u2}}$$

$$\tan(35^0) = \frac{5.493}{27.49 - v_{u2}}$$

$$v_{u2} = 19.645 \text{ m/s}$$

The work done by impeller per kg of water per second is given by

$$= \frac{1}{g} v_{u2} u_2$$

$$= \frac{1}{9.81} \times 19.645 \times 27.49$$

$$= 55.05 \frac{\text{Nm}}{\text{N}}. \textbf{ Ans.}$$

EXAMPLE 4.1

4.3 EFFICIENCY OF CENTRIFUGAL PUMPS

Efficiency of a pump can be expressed into three different ways as;

(i) Mechanical efficiency

(ii) Hydraulic efficiency

(iii) Volumetric efficiency

4.3.1 Mechanical Efficiency

As discussed earlier that pump converts the mechanical power into hydraulic power. The mechanical power from the motor or other driver is transferred to the shaft of pump through coupling or transmitting mechanism. There are some losses occur between the driver and pump which include bearing losses, sealing losses and power transmission losses from driver shaft to pump shaft. It is therefore, the power consumed by the pump impeller is lesser than the power transmitted by the motor or driver. Considering these mechanical losses, the pump efficiency is expressed in terms of mechanical efficiency as given by the following expression

$$\eta_{mechanical} = \frac{P_{impeller}}{P_{shaft}} \tag{4.14}$$

4.3.2 Hydraulic Efficiency

In order to create the pressure of fluid in the form potential energy, pump consumes mechanical power supplied by the driver. During this process, some losses occur over the impeller in the form of hydraulic losses and the corresponding efficiency is expressed in terms of hydraulic efficiency as

$$\eta_{hydraulic} = \frac{\rho g Q H}{P_{impeller}} \tag{4.15}$$

Where ρ is density of water, g is acceleration due to gravity, Q is discharge of water and H is head.

4.3.3 Volumetric Efficiency

During operation of a pump, it is expected that the amount of fluid received by the pump at its suction side should be delivered in the same quantity at its delivery side. However, in actual practice there is always some loss of fluid in terms of leakage inside the pump. It means that $Q_{inlet} > Q_{outlet}$ and due to this loss of fluid amount, the pump efficiency is expressed as volumetric efficiency as

$$\eta_{volumetric} = \frac{Q_{total} - Q_{leakage}}{Q_{total}} \tag{4.16}$$

Where, Q_{total} is total discharge, $Q_{leakage}$ is discharge losses inside the pump, and Q_{total} is total discharge.

Considering all the three efficiencies together, the pump efficiency is expressed in terms of overall or total efficiency as

$$\eta_{total} = \eta_{mechanical} \times \eta_{hydraulic} \times \eta_{volumetric} \tag{4.17}$$

Finally, the total efficiency can be expressed as;

$$\eta_{total} = \frac{\rho g Q H}{P_{shaft}} \tag{4.18}$$

SCAN ME

For better understanding pump efficiency

Example 4.2 A pump requires a power of 100 W to pump the water discharge of 1.10 m^3/h against a head of 12 m. Determine the overall efficiency of the pump.

Solution

Given

Power, P = 100 W

Discharge, Q = 1.10 m^3/h = 3.05 x 10^{-4} m^3/sec

Head, H = 12 m

Overall efficiency,

$$\eta_{total} = \frac{\rho g H Q_{net}}{P_{shaft}}$$

$$= \frac{1000 \times 9.81 \times 12 \times 3.05 \times 10^{-4}}{100}$$

$$= 0.8677 \times 100$$

η_{total} = 86.77 %. **Ans.**

EXAMPLE 4.2

Example 4.3 Determine the inlet tip angle for centrifugal pump impeller whose outer diameter is double of inner diameter and speed is 1500rpm. Consider the total head of pump as 45m and constant flow velocity over Impeller as 2.25 m/s. Other dimensions are as outer diameter as 525mm, outlet width of 45mm and vane outlet tip angle as 40°. Also find out the manometric efficiency and work done by the impeller for unit discharge.

Solution

Given data

N = 1500 rpm

H_m = 45 m

$v_{f_1} = v_{f_2}$ = 2.25 m/s

β_1 = 40°

D_2= 525 mm= 0.525 m

B_2 = 45 mm = 0.045 m

Inner dia. of impeller, is given by

$$D_1 = \frac{D_2}{2} = \frac{0.525}{2} = 0.2625 \text{ m}$$

Tangential velocity of impeller at inlet and outlet are

$$u_1 = \frac{\pi D_1 N}{60} = \frac{\pi \times 0.2625 \times 1500}{60} = 20.61 \text{ m/s}$$

and

$$u_2 = \frac{\pi D_2 N}{60} = \frac{\pi \times 0.525 \times 1500}{60} = 41.23 \text{ m/s}$$

EXAMPLE 4.3

EXAMPLE 4.3

Discharge is given by,

$Q = \pi D_2 B_2 \times V_{f_2} = \pi \times 0.525 \times .045 \times 2.25 = 0.167\ \text{m}^3/\text{s}.$

From inlet velocity triangle

$$\tan \alpha_1 = \frac{v_{f_1}}{u_1} = \frac{2.25}{20.61} = 0.109$$

$\therefore\ \alpha_1 = \tan^{-1} 0.109 = 6.23°$ Ans.

But from outlet velocity triangle, we have

$$\tan \beta_1 = \frac{v_{f_2}}{u_2 - v_{u_2}} = \frac{2.25}{(41.23 - v_{u_2})}$$

$$\tan 40^o = \frac{v_{f_2}}{u_2 - v_{u_2}} = \frac{2.25}{(41.23 - v_{u_2})}$$

$$v_{u2} = 38.54\ \text{m/s}$$

The work done by impeller is given by

$$= \frac{W}{g} \times v_{u_2} u_2 = \frac{\rho \times g \times Q}{g} \times v_{u_2} \times u_2$$

$$= \frac{1000 \times 9.81 \times 0.167}{9.81} \times 38.54 \times 41.23$$

$$= 265363.70\ \text{Nm/s.}\ \textbf{Ans.}$$

Manometric efficiency (η_{man}) is given by

$$\eta_{man} = \frac{gH_m}{v_{u2} u_2}$$

$$\eta_{man} = \frac{9.81 \times 45}{38.54 \times 41.23}$$

$\eta_{man} = 0.2778 = 27.78\%.$ **Ans.**

EXAMPLE 4.4

Example 4.4 A pump is installed having the following parameters as: (i) impeller diameter: 410 mm (ii) speed: 750 r.p.m (iii) head: 12 m (iv) vane angle: 42^0 (v) width: 50 mm and (vi) manometric efficiency: 80%. Find out flow velocity at outlet, water velocity leaving the vane, angle of absolute velocity at outlet and discharge.

Solution.

Given data:

$D_2 = 410\ \text{mm} = 0.41\ \text{m}$

$B_2 = 50\ \text{mm} = 0.05\ \text{m}$

$N = 750\text{ r.p.m}$

$H_m = 12\text{ m}$

$\eta_{man} = 80\% = 0.8$

$\beta_1 = 42°$

Tangential velocity of impeller is given by

$$u_2 = \frac{\pi D_2 N}{60} = \frac{\pi \times 0.41 \times 750}{60} = 16.10\text{ m/sec}$$

Manometric efficiency is given by

$$\eta_{man} = \frac{gH_m}{v_{u_2} u_2}$$

$$0.80 = \frac{9.81 \times 12}{v_{u_2} \times 16.10}$$

$$v_{u_2} = \frac{9.81 \times 12}{0.8 \times 16.10} = 9.139\text{ m/s.}$$

From the outlet velocity triangle, we have

$$\tan\beta_2 = \frac{v_{f_2}}{u_2 - v_{u_2}} = \frac{v_{f_2}}{(16.10 - 9.139)} = \frac{v_{f_2}}{6.961}$$

$\therefore \quad v_{f_2} = 6.961 \tan\beta_2 = 6.961 \times \tan 42° = 4.267\text{ m/s.}$**Ans.**

Velocity of water leaving the vane (v_2) is given by

$$v_2 = \sqrt{v_{f_2}^2 + v_{u_2}^2} = \sqrt{4.267^2 + 9.139^2}$$

$$= \sqrt{18.2 + 83.52} = 10.08\text{ m/s.}$$ **Ans.**

Angle made by absolute velocity at outlet (β_2), is given by

$$\tan\beta_2 = \frac{v_{f_2}}{v_{u_2}} = \frac{4.267}{9.139} = 0.467$$

$\therefore \quad \beta_2 = \tan^{-1} 0.467 = 25.02°$.**Ans.**

Discharge through pump is given by,

$Q = \pi D_2 B_2 \times v_{f_2} = \pi \times 0.41 \times 0.05 \times 4.267 = 0.274\text{ m}^3\text{/s.}$ **Ans.**

EXAMPLE 4.4

Example 4.5 Determine the power required to drive a centrifugal pump to deliver 0.035 m³/s of water against a head of 25 m through a 16 cm diameter pipe and 150 m long. The overall efficiency of the pump is 75% Use the the formula $h_f = \frac{4fLV^2}{2gd}$ to determine the frictional losses in the pipe having co-efficient of friction 'f' as 0.15.

Solution:

Given data:

$Q = 0.035\ \text{m}^3/\text{s}$

$H_m = h_s + h_d = 25\ \text{m}$

$D_s = D_d = 16\ \text{cm} = 0.16\ \text{m}$

$L = L_s + L_d = 150\ \text{m}$

$\eta_o = 75\% = 0.75$

$f = 0.15$

Velocity of water in pipe, is given by

$$v_s = v_d = v = \frac{\text{Discharge}}{\text{Area of pipe}}$$

$$v_s = \frac{0.035}{\frac{\pi}{4} \times 0.16^2} = 1.74 \frac{\text{m}}{\text{s}}$$

Frictional head loss in pipe, is given by

$$\left(h_{f_s} + h_{f_d}\right) = \frac{4fLV^2}{d \times 2g} = \frac{4 \times 0.015 \times 150 \times 1.74^2}{0.16 \times 2 \times 9.81} = 8.68\ \text{m.}\ \textbf{Ans}$$

Monomeric head is given by

$$H_m = (h_s + h_d) + \left(h_{f_s} + h_{f_d}\right) + \frac{v_d^2}{2g}$$

$$= 25 + 8.68 + \frac{1.74^2}{2 \times 9.81}$$

$$= 33.83\ \text{m}$$

Overall efficiency is given by

$$\eta_o = \frac{\rho g \times Q \times H_m}{1000 \times \text{S.P.}}$$

Shaft power is given by

$$\text{Shaft power} = \frac{\rho g \times Q \times H_m}{1000 \times \eta_o}$$

$$\text{P} = \frac{1000 \times 9.81 \times .035 \times 33.83}{1000 \times 0.75}$$

P= 15.487 kW. **Ans.**

EXAMPLE 4.5

Example 4.6 Find out the head and power developed to deliver a discharge of 60 lps by a three-stage centrifugal pump having impeller diameter, width and speed as 350 mm, 25 mm and 750 r.p.m respectively. The vanes having an angle 45^0 and occupy 10% of circumferential area. Assume the manometric efficiency and the overall efficiency as 90% and 80% respectively.

Solution

Given data,

n = 3

$Q = 0.06\ m^3/s$

$D_2 = 35\ cm = 0.35\ m$

$B_2 = 25\ mm = 0.025\ m$

$N = 750\ rpm$

$\beta_1 = 45^0$

$\eta_o = 80\% = 0.8$

$\eta_m = 90\% = 0.9$

Reduction in area at outlet = 10% = 0.1

Area of flow at outlet = $0.9 \times \pi \times D_2 \times B_2$

$= 0.9 \times \pi \times 0.35 \times 0.025$

$= 0.0247\ m^2$

Velocity of flow at outlet, is given by

$$v_{f_2} = \frac{\text{Discharge}}{\text{Area of flow}} = \frac{0.06}{0.0247} = 2.43\ \text{m/s}$$

Tangential velocity of impeller at outlet, is given by

$$u_2 = \frac{\pi D_2 N}{60} = \frac{\pi \times 0.35 \times 750}{60} = 13.744\ \text{m/s}$$

Refer to Fig. From velocity triangle at outlet,

$$\tan \beta_1 = \frac{v_{f_2}}{u_2 - v_{u_2}}$$

$$u_2 - v_{u_2} = \frac{v_{f_2}}{\tan \beta_1} = \frac{2.43}{\tan 45°} = 2.43\ \text{m/s}$$

$$v_{u_2} = u_2 - 2.43 = 13.744 - 2.43 = 11.314\ \text{m/s}$$

Manometric efficiency is given by

$$\eta_{\text{man}} = \frac{gH_m}{V_{u_2} u_2}$$

EXAMPLE 4.6

$$0.90 = \frac{9.81 \times H_m}{11.314 \times 13.744}$$

$$H_m = \frac{0.90 \times 11.534 \times 13.744}{9.81} = 14.54 \text{ m}$$

Total head generated by pump is given by

$$= n \times H_m = 3 \times 14.54 = 43.63 \text{ m. } \textbf{Ans.}$$

Power output of the pump is given by

$$P = \frac{\text{Weight of water lifted } \times \text{ Total head}}{1000}$$

$$P = \frac{\rho g \times Q \times 43.63}{1000}$$

$$P = \frac{1000 \times 9.81 \times 0.06 \times 43.63}{1000}$$

$$= 25.68 \text{ kW}$$

Overall efficiency (η_o) is given by

$$\eta_o = \frac{\text{Power output of pump}}{\text{Power input to the pump}} = \frac{25.68}{\text{Shaft power}}$$

$$\text{Shaft power} = \frac{25.68}{\eta_o} = \frac{25.68}{0.80} = 32.1\text{kW. } \textbf{Ans.}$$

EXAMPLE 4.6

4.4 SERIES AND PARALLEL OPERATION OF CENTRIFUGAL PUMPS

4.4.1 Pumps in Series

As we know that centrifugal pumps convert mechanical power into potential power in terms of head and discharge of a liquid. The head(pressure) developed by the pump depends upon the speed of the pump and diameters of impeller. For a given requirement (head and discharge), the pump parameters (pump speed, impeller diameters and breadth) are designed for best efficiency point. For design point of view, there are constraints of increasing head developed by a pump beyond its design head. If more head is needed for a given discharge, two or more pumps can be connected in series. The higher heads may also be produced by using multi-stage pumps, where, two or more impeller are fitted on the same shaft in series in a casing as shown in Fig. 4.9 for a two-stage pump. In the case of multi-stage pumps, the pressure (head) of a pump is added in the head developed by the next pump for a given quantity of liquid and so on. According to the number of impellers fitted in the casing a multi-stage pump is designated as two stage, three-stage, etc.

If two pumps or impellers of two-stage pump having head as H_1 and H_2, then total head developed by the pumps in series for the given discharge will be as $H_m = (H_1 + H_2)$. If there are n pumps having respective head as; $H_1, H_2, H_3, \ldots\ldots H_n$, then the total head will be as

$$H_t = H_1 + H_2 + H_3 + \cdots + H_n \tag{4.19}$$

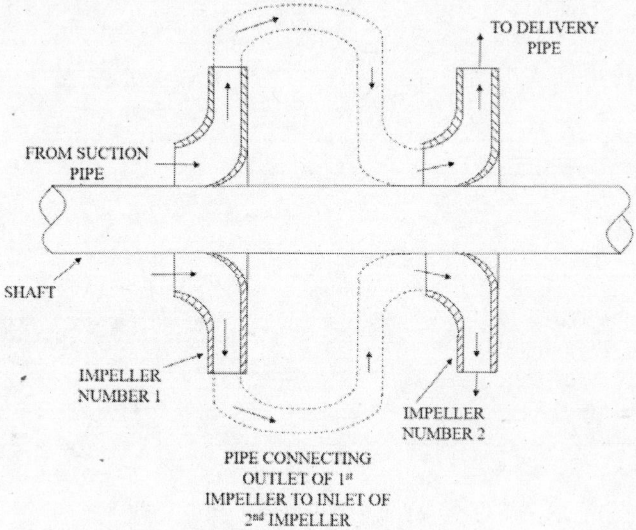

Fig. 4.9: Two-stage centrifugal pump

4.4.2 Pumps in Parallel

As discussed above, to create more head, the pumps can be connected in series. In case of more discharge requirement by the pump where a single pump is not capable to deliver required discharge, then number of pumps can be connected in parallel, as shown in Fig. 4.10.
As the pumps having the same head and connected in parallel, the total head will be the same as of a single pump. However, the total discharge will be added and can be expressed as

$$Q_t = Q_1 + Q_2 + Q_3 \ldots\ldots\ldots\ldots Q_n \tag{4.20}$$

The Eq. 4.20 can be written as follows if n number of pumps having the same discharge Q and connected in parallel.

$$Q_t = nQ \tag{4.21}$$

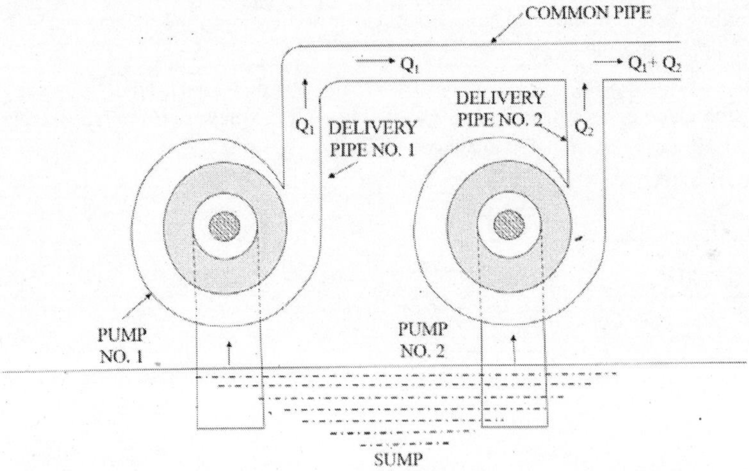

Fig. 4.10: Two centrifugal pumps arranged in parallel

EXAMPLE 4.7

Example 4.7 Determine the number of the pumps required in series to deliver a discharge of 65 liters/s against a total head of 180 m, at a speed of 1000 r.p.m. The specific speed is not to exceed 750.

Solution

Given data:

N_S=750

Q= 65 l/s

H=180 m

N=1000 rpm

Specific speed is given by

$$N_s = \frac{N\sqrt{Q}}{H_m^{3/4}}$$

$$750 = \frac{1000(65)^{1/2}}{H_m^{3/4}}$$

$$H_m = 23.72 \text{ m}$$

$$\text{Head per stage} = H_m = 23.72 \text{ m}$$

Number of stages $= \frac{180}{23.72} = 7.58 = 8$. **Ans.**

Example 4.8 Determine the number of pumps each having a discharge capacity of 120 liters/s to deliver this discharge against a head of 100m. The specific speed of the pump should not exceed 850 for given speed of pump as 1000 rpm.

Solution

Given data:

$N_s = 850$

$Q = 120$ l/s

$H = 100$ m

N = 1000 rpm

Specific speed is given by

$$N_s = \frac{N\sqrt{Q}}{H_m^{3/4}}$$

$$850 = \frac{1000(120)^{1/2}}{H_m^{3/4}}$$

$$H_m = 30.21 \text{ m}$$

Number of pumps needed $= \frac{100}{30.21} = 3.30 = 4$. **Ans.**

Since the total head required to be developed is more than the head developed by each pump, the pumps should be connected in series.

EXAMPLE 4.8

Example 4.9 Determine the efficiency of a four-stage centrifugal pump which is running at 500 r.p.m. to create a total head of 45 m to discharge water of 0.15 m^3/s. If the width and diameter of each impeller at outlet is 6 cm and 70 cm respectively having vane tip angle of 40°. Find the manometric efficiency.

Solution

Given data:

$n = 4$

$N = 500$ rpm

$H_m = 45$ m

$Q = 0.15\ m^3/s$

$\beta_1 = 40^0$

$B_2 = 6$ cm $= 0.06$ m

$D_2 = 70$ cm $= 0.7$ m

Head per stage is given by

$$= \frac{45}{4} = 11.25 \text{ m}$$

Tangential velocity of impeller at outlet, is given by

$$u_2 = \frac{\pi D_2 N}{60} = \frac{\pi \times 0.7 \times 500}{60} = 18.32 \frac{\text{m}}{\text{s}}$$

EXAMPLE 4.9

EXAMPLE 4.9

Velocity of flow at outlet, is given by

$$v_{f_2} = \frac{\text{Discharge}}{\text{Area of flow}} = \frac{0.15}{\pi D_2 B_2} = \frac{0.20}{\pi \times 0.7 \times 0.06} = 1.515 \text{ m/sec}$$

From velocity triangle at outlet,

$$\tan \beta_1 = \frac{v_{f2}}{u_2 - v_{u_2}}$$

$$u_2 - v_{u_2} = \frac{v_{f2}}{\tan \beta_1} = \frac{1.515}{\tan 40^o} = 1.806 \text{ m/sec}$$

$$v_{u_2} = u_2 - 1.806 = 18.32 - 1.806 = 16.514 \text{ m/s}$$

Manometric efficiency is given by

$$\eta_{man} = \frac{gH_m}{V_{u_2} u_2} = \frac{9.81 \times 11.25}{16.514 \times 18.32}$$

$$\eta_{man} = 0.364 \text{ or } 36.4\ \%.$$ **Ans.**

4.5 PERFORMANCE TESTING OF PUMP

As centrifugal pumps are designed for a desired operating range of head (H) and discharge (Q) which is inherently a low range. It is therefore, the operating performance of a pump should be known by conducting the tests in the laboratory. The test results are then used to represent the performance of the pump in the form of performance characteristics curves. A schematic of the test rig for conducting tests of a pump is shown in Fig. 4.11.

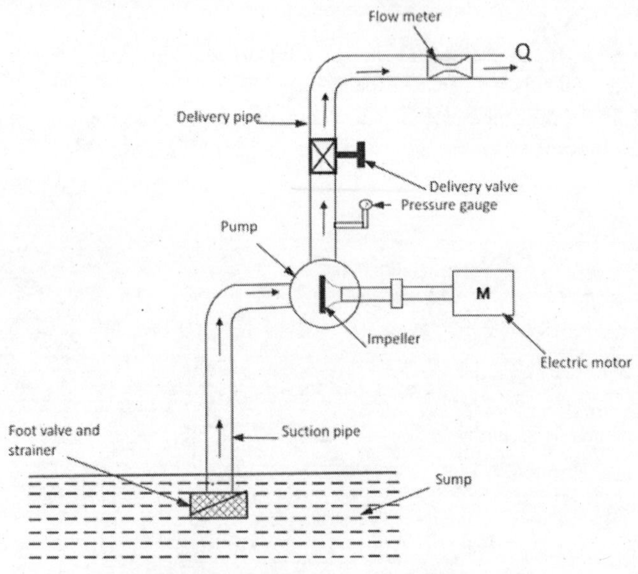

Fig. 4.11: Test rig for conducting test of pump

Typical test rig for conducting performance test of a pump consists of (i) test pump (ii) water sump (iii) driver (electric motor) (iv) control valve (v) flow meter (vi) pressure gauge (vii) tachometer (viii) power measuring instrument for input power of motor.

4.5.1 Procedure

The performance tests on the pumps are carried out by following the steps given below.

Step-1: After priming the pump the delivery valve is fully closed and pump is started at rated speed by driver (electric motor).

Step-2: At this condition pressure gauge readings are taken for head. However, the discharge will be zero. The developed pressure head is known as 'shut-off head' of the pump.

Step-3: Now the delivery valve is partially opened and readings for all the parameters such as pressure gauge readings, flow meter, torque meter (T) and speed (N) are taken.

Step-4: Step-3 is repeated for different opening of delivery valve (from closed position to fully open positions) for number of test runs.

Step-5: The efficiency (η) of the pump at the different operating points is calculated considering the reading taken and using the following formulae

$$\eta = \frac{\text{Hydraulic power}}{\text{Input power}}$$

$$\eta = \frac{\rho g Q H}{T\omega} \tag{4.22}$$

Where, ρ is density of water (kg/m^3), g is acceleration due to gravity (m^2/sec), Q is discharge (m^3/sec), H is head (m), T is torque (N-m), ω is angular speed of shaft (rad/sec).

$$\omega = \frac{2\pi N}{60} \tag{4.23}$$

Where, N is speed (r.p.m.)

The calculated values for various parameters are presented in Table 4.1.

Table: 4.1: Determined values for various parameters of testing pump

Parameters	**Opening of the control valve**				
	0%	**25%**	**50%**	**75%**	**100%**
Discharge, Q					
Head, H					
Input power					
Hydraulic power					
Efficiency					

Using the values listed in Table 4.1, following curves are drawn to represent the performance of a pump.

(i) Head (H) versus discharge (Q)
(ii) Input power (P) versus discharge (Q)
(iii) Efficiency (η) versus discharge (Q)

Fig. 4.12 shows the performance curves for a typical centrifugal pump.

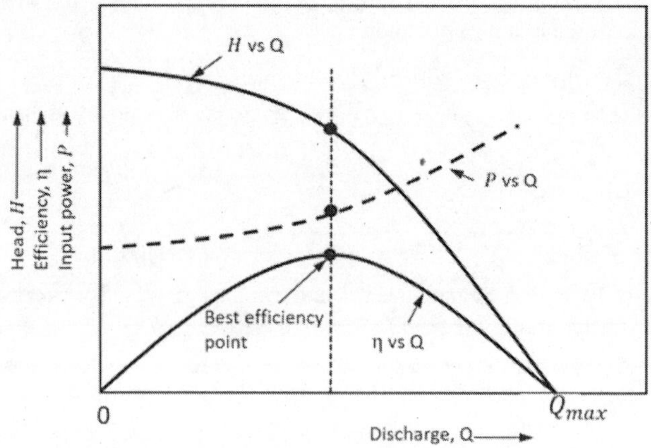

Fig. 4.12: Performance curves for a typical centrifugal pump

4.6 PERFORMANCE CURVES

Basically, pumps are designed for a given requirement of head H and discharge Q accordingly their efficiency is determined and specified for a single point of operation. However, in actual practice the pump may be operated at various points. The operating behavior of the pump is necessary to know for its operation at different points which can be represented by three different types of characteristic curves. Following are the characteristic curves which represents the performance behavior of a pump:

(i) Main characteristics curves
(ii) Operating characteristics curves
(iii) Constant efficiency curves

4.6.1 Main Characteristics Curves

While designing a pump, designer wants to know the behaviour of pumps in terms of a particular parameter with respect to given parameters. The behaviour of pump under such conditions is represented by main characteristic curves. As shown in Fig. 4.13, the head decreases as the discharge increases for a given speed. The similar trends will be followed for different speeds which is shown in discharge Q versus head H curves. In case of input power with respect to discharge it increases with discharge. However, for a given speed the efficiency of the pump initially increases with discharge, achieves a maximum efficiency and then declines as shown in discharge, Q versus efficiency, η curves.

In some pumping system discharge and head need to be varied for an installing pump which can be done by varying the speed of the pump. Based on the characteristics of the pump, discharge is proportional to speed, head is proportional to square of speed and power is proportional to cubic of speed. For such cases the characteristic curves as shown in Fig. 4.14 are prepared.

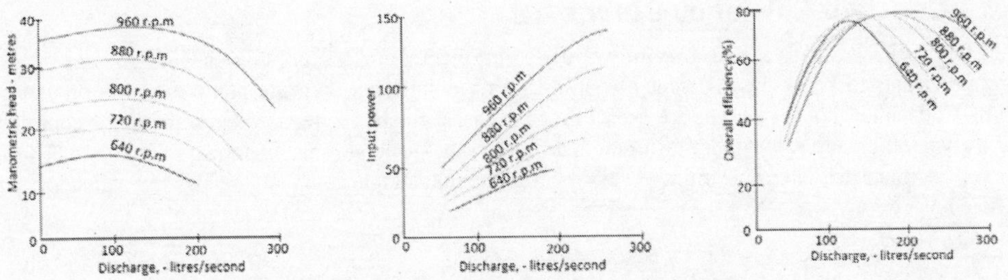

Fig. 4.13: Main characteristics of a centrifugal pump

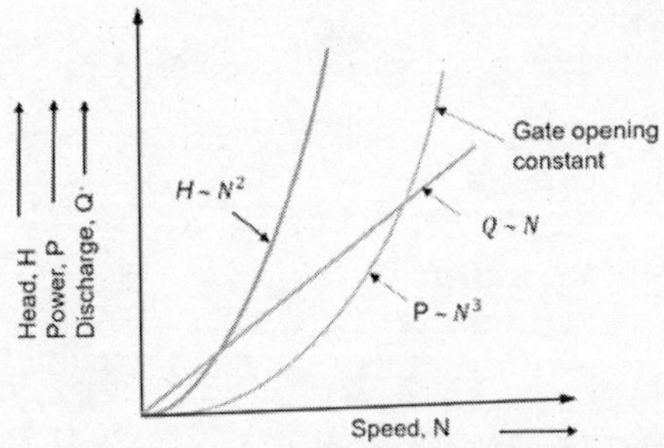

Fig. 4.14: Q, P, H versus N curves of a centrifugal pump.

SCAN ME

For better understanding performance curves

4.6.2 Operating Characteristics Curves

Further, for practical applications a pump normally is operated at constant speed for a design head and discharge. However, pump discharge can be varied according to the user's requirement. Under this condition of pump operation, the head will also vary due to change in frictional losses in piping system. The characteristic of pump under these conditions is represented by operating characteristic curves as shown in Fig. 4.15.

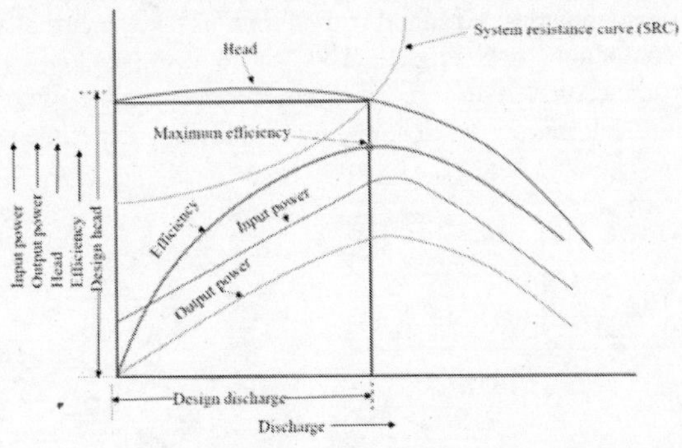

Fig. 4.15: Operating characteristics curves of a centrifugal pump

4.6.3 Constant Efficiency Curves

Constant efficiency curves are also known as iso-efficiency or Muschel curves. In some cases, the operator wants to know the efficiency at an operating point other than the pump's design operating point. Constant efficiency curves are useful for such cases which are prepared by drawing the contours for a given efficiency value under different head and discharge conditions, as shown in Fig. 4.16. These curves facilitate to determine number of operating points for a pump at the desired value of the efficiency.

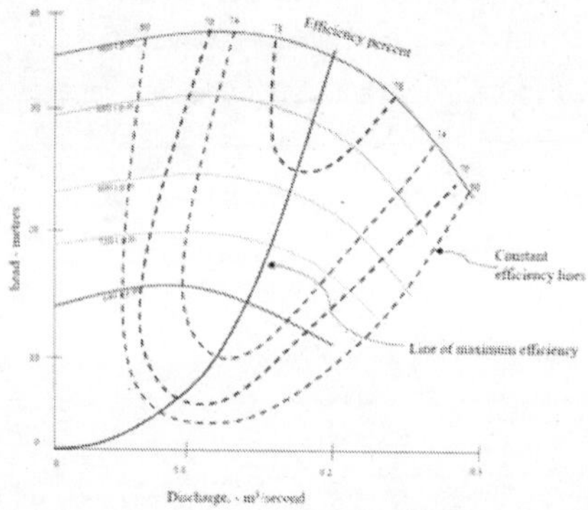

Fig. 4.16: Constant efficiency of a centrifugal pump

4.7 CAVITATION IN PUMPS

Cavitation is an undesired condition in centrifugal pump operation which should be avoided. A pump is operated under positive pressure at delivery side and under negative pressure at suction side. If the pressure at the suction side drops below the vapour pressure of the liquid, then the vapour formation takes place at normal temperature of liquid. The vapour in the form of bubbles is carried with the water flow from low pressure region to high pressure region inside the pump. These bubbles in high pressure region are collapsed and very high velocity jet with very high frequency strikes over the surface of the pump component. Due to which undesired sound with vibrations take place and after due course of time the material from the component gets eroded. This phenomenon is known as "Cavitation". The cavitation in a pump cause the sudden drop in efficiency and reduces the life of the pump. In order to avoid cavitation, a factor determined by Thoma by conducting experiments which is known as "Thoma's coefficient" or "Thoma's factor" and expressed as

Thoma's cavitation factor (σ),

$$\sigma = \frac{\left[\frac{p_a - p_v}{w}\right] - (h_s + h_{fs})}{H_m} \tag{4.24}$$

Where,

p_a is Atmospheric pressure

p_v is Vapour pressure of water

w is specific weight

h_s is suction head

h_{fs} is head due to friction losses

H_m is manometric head

SCAN ME

For better understanding cavitation in pumps

Cavitation factor may also be expressed in terms of Net Positive Suction Head (NPSH), which can be defined as; head required to make the liquid to flow through the suction pipe to the impeller.

$$\sigma = \frac{\text{NPSH}}{H_m} \tag{4.25}$$

To avoid the cavitation, the critical value of Thoma's coefficient should be determined. It can be defined as the value of cavitation factor below which the cavitation will likely to occur. It is generally determined during testing of pump in the laboratory. However, it can be determined by using the following empirical relation

$$\sigma_c = 0.103\left(\frac{N_s}{1000}\right)^{\frac{4}{3}} \tag{4.26}$$

Where, N_s is specific speed.

Example 4.10 Calculate the net power suction head (NPSH) for a centrifugal pump having speed of 750 r.p.m. delivers 0.15 cumec. of water against a head of 25m. Take the atmospheric pressure as 1×10^5 Pa(abs.) and vapour pressure of water as 3 kPa (abs.). Consider the frictional losses in the pipe as 0.15m.

Solution

Given data:

$N = 750$ r.p.m

$Q = 0.15\ \text{m}^3/\text{s}$

$H_m = 25$ m

$P_a = 1 \times 10^5\ \text{Pa} = 1 \times 10^5\ \text{N/m}^2$

$P_v = 3\ \text{kPa} = 3 \times 10^3\ \text{N/m}^2$

Critical value of σ i.e., σ_c is given by

$$\sigma_c = 1.03 \times 10^{-3} \times N_s^{4/3}$$

$$N_s = \text{specific speed of pump} = \frac{N\sqrt{Q}}{H_m^{\frac{3}{4}}}$$

$$N_s = \frac{750 \times \sqrt{0.15}}{25^{3/4}} = 25.98$$

$$\sigma_c = 1.03 \times 10^{-3} \times 25.98^{4/3} = 0.07925$$

EXAMPLE 4.10

EXAMPLE 4.10

Thoma's cavitation factor (σ), is given by

$$\sigma = \frac{NPSH}{H_m}$$

From the above equation, NPSH is directly proportional to Thoma's cavitation factor (σ). NPSH will be minimum when σ is minimum. But the minimum value of σ for no cavitation is σ_c .Hence when $\sigma = \sigma_c$ then NPSH is minimum.

$$\therefore \quad \sigma_c = \frac{(NPSH)_{min}}{H_m}$$

or, $\quad (NPSH)_{min} = H_m \times \sigma_c$

$(NPSH)_{min} = 0.07925 \times 25 = 1.98$ m. **Ans**

EXAMPLE 4.11

Example 4.11 Determine the suction head for a centrifugal pump installed to avoid cavitation for the given parameters of; atmospheric pressure 100 kPa; vapour pressure 2.30 kPa; inlet and other losses in suction pipe 1.50m; effective head of pump 52.0 m; and cavitation parameter $\sigma = 0.117$.

Solution

Given data,

P_a = 100kPa=100000 N/m^2

P_v = 2.30 kPa=2300 N/m^2

H_m = 52 m

h_{fs} =1.5 m

σ = 0.117

Cavitation factor is given by

$$\sigma = \frac{[(p_a - p_v)/w] - (h_s + h_{fs})}{H_m}$$

Thus, by substitution, we get

$$0.117 = \frac{[(100000 - 2300)/9810] - (h_s + 1.5)}{52}$$

$$0.117 \times 52 = \frac{(100000 - 2300)}{9810} - h_s - 1.55$$

$$\therefore \quad h_s = 2.325 \text{ m } \textbf{Ans}$$

i.e., the pump maybe installed at a height of 2.325 m above the water surface in the supply tank.

4.8 MODEL TESTING OF CENTRIFUGAL PUMPS

To validate the efficiency of a pump prototype, it is not possible to test the prototype of large capacity pump in the laboratory, it is therefore, model testing is performed based on similarity law. Similarity law states that two machines of different capacity can said similar if they are geometrically, kinetically and dynamically similar. Based on the similarity law, a term known as specific speed, N_s is derived. In case of turbine, the specific speed is expressed in terms of power, while in case of pumps it is expressed in terms of discharge.

4.8.1 Specific Speed of Pumps

Specific speed (N_s) of a pump can be defined as the speed of pump to deliver unit discharge against a unit head.

The derivation for a specific speed is derived as follows as

The discharge, Q for a centrifugal pump is given by the relation

$$Q = \text{Area} \times \text{Velocity of flow}$$

$$= \pi D \times B \times v_f$$

$$Q \propto D \times B \times v_f \tag{4.27}$$

Where, D is diameter of the impeller of the pump and B is width of the impeller.

We know that $B \propto D$

∴ From equation (4.27), we get

$$Q \propto D^2 \times v_f \tag{4.28}$$

The tangential velocity is given by

$$u = \frac{\pi DN}{60} \propto DN \tag{4.29}$$

Now the tangential velocity (u) and velocity of flow (v_f) are related to the manometric head (H_m) as

$$u \propto v_f \propto \sqrt{H_m} \tag{4.30}$$

Compare the Eq. 4.29 and Eq. 4.30, we get

$$\sqrt{H_m} \propto DN$$

$$\text{or } D \propto \frac{\sqrt{H_m}}{N}$$

Substituting the values of D in Eq. (4.28)

$$Q \propto \frac{H_m}{N^2} \times v_f$$

$$\propto \frac{H_m}{N^2} \times \sqrt{H_m} \ \left[\text{where, } v_f \propto \sqrt{H_m}\right]$$

$$\propto \frac{H_m^{3/2}}{N^2}$$

$$Q = \text{K}\frac{H_m^{3/2}}{N^2} \tag{4.31}$$

Where, K is a constant of proportionality.

If $H_m = 1$ m and $Q = 1\ \text{m}^3/\text{s}$, N becomes N_s.

Substituting these values in Eq. (4.31), we get

$$1 = K\frac{1^{3/2}}{N_s^2} = \frac{K}{N_s^2}$$

$$\mathrm{K} = N_s^2$$

$$Q = N_s{}^2\frac{H_m^{3/2}}{N^2}$$

$$N_s^2 = \frac{N^2 Q}{H_m^{3/2}}$$

$$N_s = \frac{N\sqrt{Q}}{H_m^{3/4}} \tag{4.32}$$

The complete similarity between the model and actual pump (prototype) will exist if the following condition are satisfied:

(i) Specific speed of model = Specific speed of prototype

$$(N_s)_m = (N_s)_p$$

$$\left(\frac{N\sqrt{Q}}{H_m^{3/4}}\right)_m = \left(\frac{N\sqrt{Q}}{H_m^{3/4}}\right)_p \tag{4.33}$$

(ii) Tangential velocity (u) is given by $u = \frac{\pi DN}{60}$ also $u \propto \sqrt{H_m}$

$$\sqrt{H_m} \propto DN$$

$$\frac{\sqrt{H_m}}{DN} = \text{Constant}$$

$$\left(\frac{\sqrt{H_m}}{DN}\right)_m = \left(\frac{\sqrt{H_m}}{DN}\right)_p \tag{4.34}$$

(iii) Discharge

$$Q \propto D^2 \times V_f$$
$$\propto D^2 \times D \times N$$
$$\propto D^3 \times N$$

Where,

$$v_f \propto u \propto DN$$

$$\frac{Q}{D^3 N} = \text{constant}$$

$$\text{or } \left(\frac{Q}{D^3N}\right)_m = \left(\frac{Q}{D^3N}\right)_p \tag{4.35}$$

(iv) Power of the pump

$$P = \frac{\rho g Q H_m}{75}$$

$$P \propto Q \times H_m$$

$$\propto D^3 \times N \times H_m \; (\because\; Q \propto D^3 N \text{ and } \sqrt{H}_m \propto DN)$$

$$\propto D^3 N \times D^2 N^2$$

$$\propto D^5 N^3$$

$$\frac{P}{D^5 N^3} = \text{Constant}$$

$$\left(\frac{P}{D^5 N^3}\right)_m = \left(\frac{P}{D^5 N^3}\right)_p \qquad (4.36)$$

Example 4.12 Find out the specific speed of a pump for a given discharge of 2 m³/sec and head of 20 m if the speed of the pump 3000 r.p.m. With the suitable assumptions, find out the discharge, input power and head of the pump having 2400 r.p.m assuming the efficiency of pump as 80%.

Solution

Given data,

N = 3000 r.p.m

Q = 2 m^3/s

H = 20 m

η_o = 80 %=0.8

Power is given by

$$P = \frac{wQH}{\eta_0}$$

$$P = \frac{9810 \times 2 \times 20}{0.8}$$

$$P = 490500W = 490.5 \text{ kW}$$

The specific speed is given by,

$$N_s = \frac{N\sqrt{Q}}{H_m^{3/4}}$$

$$Q = 2\frac{m^3}{s} = 2 \times 10^3 \text{ l/s}$$

However, if discharge Q is taken in m^3/s

$$N_s = \frac{3000\sqrt{2}}{20^{\frac{3}{4}}} = 448.6$$

EXAMPLE 4.12

EXAMPLE 4.12

Since Q is proportional to N ; H is proportional to N^2 and P is proportional to N^2, we have

$$\frac{Q}{N}=\frac{Q_1}{N_1}$$

$$Q_1=\left(\frac{N_1}{N}\right)Q$$

$$Q_1=\frac{2400}{3000}\times 2=1.6\text{m}^3/\text{s. } \textbf{Ans.}$$

$$\frac{H}{N^2}=\frac{H_1}{N_1^2}$$

$$H_1=\left(\frac{N_1^2}{N^2}\right)H$$

$$H_1=\frac{(2400)^2}{(3600)^2}\times 18=8\text{ m. } \textbf{Ans.}$$

$$\frac{P}{N^3}=\frac{P_1}{N_1^3}$$

$$P_1=\left(\frac{N_1^3}{N^3}\right)P$$

$$P_1=\frac{(2400)^3}{(3000)^3}\times 490.5=251.136\text{ kW. } \textbf{Ans.}$$

EXAMPLE 4.13

Example 4.13 A pump having an impeller diameter of 0.25 m is installed to deliver the discharge of 30 lps to a height of 20 m. Find out head and impeller diameter of a similar pump for a discharge of 15 lps.

Solution:

Given data,

For pump No. 1,

$D_1=0.25$ m

$Q_1=0.03$ m^3/s

$H_{m1}=20$ m

For pump No. 2,

$Q_2=0.015$ m^3/s

$$\frac{N_1\sqrt{Q_1}}{H_{m_1}^{3/4}}=\frac{N_2\sqrt{Q_2}}{H_{m_2}^{3/4}}$$

$$N_1=N_2$$

$$\frac{\sqrt{Q_1}}{H_{m_1}^{3/4}}=\frac{\sqrt{Q_2}}{H_{m_2}^{3/4}}$$

$$\frac{\sqrt{.03}}{20^{3/4}} = \frac{\sqrt{.015}}{H_{m_2}^{3/4}}$$

$$H_{m_2}^{3/4} = \frac{\sqrt{.015} \times 20^{3/4}}{\sqrt{.03}} = \sqrt{\frac{.015}{.03}} \times 9.457 = 6.687$$

$$\therefore\ H_{m_2} = (6.687)^{\frac{4}{3}} = 12.6 \text{ m. } \textbf{Ans.}$$

Using Eq. (4.28),

$$\left(\frac{\sqrt{Hm}}{DN}\right)_1 = \left(\frac{\sqrt{Hm}}{DN}\right)_2$$

$$\text{or } \frac{\sqrt{Hm_1}}{D_1N_1} = \frac{\sqrt{Hm_2}}{D_2N_2}$$

$$N_1 = N_2$$

$$\frac{\sqrt{H_{m_1}}}{D_1} = \frac{\sqrt{H_{m_2}}}{D_2}$$

$$\frac{\sqrt{20}}{0.25} = \frac{\sqrt{12.6}}{D_2}$$

$$D_2 = \frac{\sqrt{12.6} \times 0.25}{\sqrt{20}} = 0.198 \text{ m} = 198 \text{ mm. } \textbf{Ans.}$$

EXAMPLE 4.13

4.9 SELECTION OF PUMPS

As pumps are designed for a particular requirement of head and discharge at given r.p.m., it is therefore, selection of pumps becomes a very important aspect for practical users. Scientifically pumps are selected based on specific speed (N_s) of a pump. For a given head, discharge and speed, the specific speed is determine using the expression given in Eq. 4.30. Based on specific speed, pumps are classified as;

(i) Radial flow pump (N_s = 10 to 40)

(ii) Mixed flow pumps with outlet edge parallel to machine axis (N_s= 40 to 80)

(iii) Mixed flow with outlet edge inclined to machine axis provided with volute chamber (N_s = 80 to 100)

(iv) High specific speed axial flow pumps delivering axially, provided with vanes (N_s = 100 to 1000)

Following are the steps to select a suitable pump for given conditions:

Step-1: For the given pumping system, determine the head losses (h_f) across the pipe and other pipe fittings correspond to the designed discharge

Step-2: Find out the net head, H_{net} (m) for the designed discharge, Q (m^3/sec) and given static head, H_s (m) as

$$H_{net} = H_s + h_f \tag{4.37}$$

Step-3: Determine the specific speed considering designed discharge, net head and speed of the pump using Eq. 4.32.

Step-4: Choose the types of pumps corresponding to the determined value of specific speed.

Step-5: If for the given parameters, the specific speed exceeds 1000, following two options need to be followed as N_s can not be more than 1000.

First option: Selected speed of the pump should be reduced to a value so that specific speed is less than 1000 for the given head and discharge.

Second option: If it is not possible to have specific speed value less than 1000 even taking the minimum possible speed of the pump then specific speed can be managed by dividing head and discharge. In this case we have to connect the number of pumps in series or parallel accordingly.

4.10 CENTRIFUGAL PUMP USED AS TURBINE (PAT)

As discussed earlier, turbine is a hydraulic machine which converts available potential energy into mechanical energy which is further used to generate electricity. Pump is a hydraulic machine which is used to convert the mechanical energy into potential energy. It is the evidence that principally pump can be used in reverse mode as turbine. It is similar to hydraulic turbine when operates in reverse mode. It can be seen from the schematic shown in Fig. 4.17 that the operation of a centrifugal pump in both the modes.

Followings are the advantages of use of pump in turbine mode;

(i) Due to the large number of standards pumps produced, standardized PAT can be significantly less expensive than a specially designed turbine.
(ii) The delivery time is generally much less for pump than for turbines.
(iii) Spare parts and maintenance or repair services all much more readily available for pumps than turbines.
(iv) The control of PAT is simple due to the absence of blade pitch change.
(v) These factors contribute to cost saving.

In spite of the advantages of pump as turbine as stated above the main problem is associated with is pump selection in turbine mode. In case of turbine operation, the operating point will shift towards higher head and discharge to achieve the same efficiency as shown in Fig. 4.18. In order to solve this problem head conversion factor (h) and discharge conversion factor (q) are determined experimentally which can be expressed as

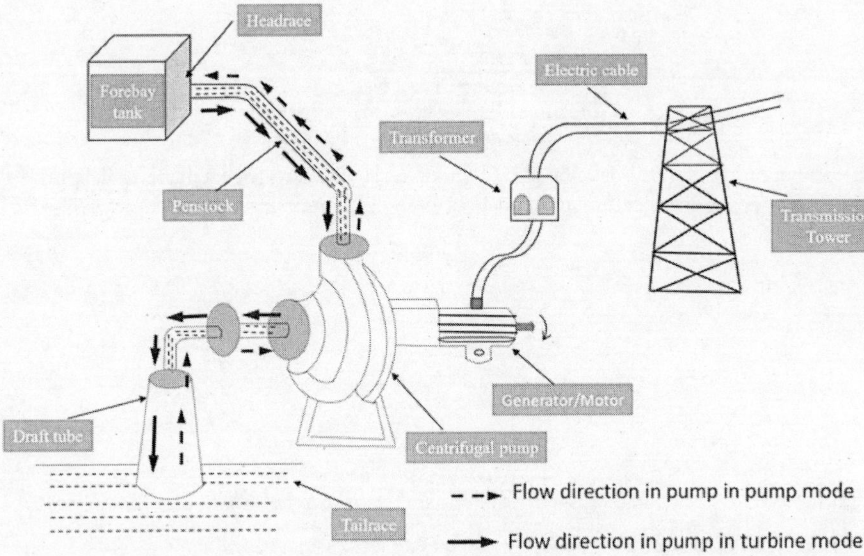

Fig. 4.17: Schematic showing centrifugal operations in pump and turbine mode

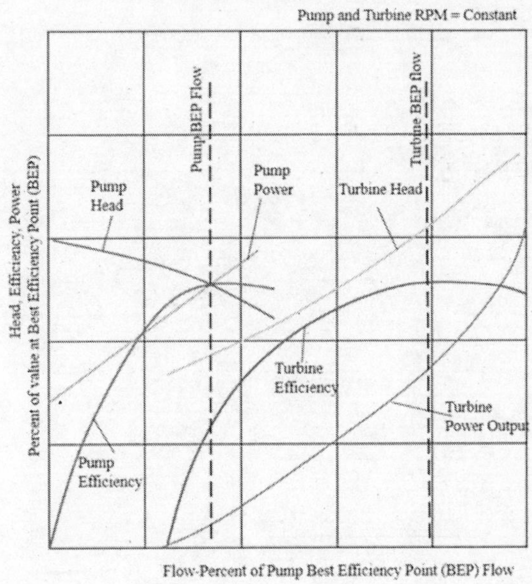

Fig. 4.18: Operating characteristic curves of a pump in both the modes (pump as pump and pump as turbine).

Conversion factor for head, $h = \frac{\text{Head in turbine mode } (H_T)}{\text{Head in pump mode } (H_P)}$

Conversion factor for discharge, $q = \frac{\text{Discharge in turbine mode } (Q_T)}{\text{Discharge in pump mode } (Q_P)}$

However, correlations have been developed by various researchers which facilitate to determine the conversion factors based on specific speed such as the correlations for these factors developed by Stepanoff's Method as

$$h = \frac{1}{\sqrt{\eta_P}} \tag{4.38}$$

$$q = \frac{1}{\sqrt{\eta_P}} \tag{4.39}$$

EXAMPLE 4.14

Example 4.14 For a pumping system, pumps are required to be designed to lift the water discharge of 0.2 $\frac{\text{m}^3}{sec}$ against a total head of 25 m. Determine the number of pumps required if each pump having capacity of 22.5 kW.
Take the other parameters as;

i. Lengths of the pipe = 200 m.
ii. Diameters of pipe = 50 cm.
iii. Friction factor = 0.01
iv. Combined efficiency of motor and pump = 80%.

Solution

Given

Pump capacity = 22.5 kW

$Q = 2.0\ \text{m}^3/\text{sec}$, $H_S = 25$ m, $L = 200$ m, $D = 0.5$ m, $f = 0.01$, $\eta_o = 0.8$

Area of pipe = $\frac{\pi}{4}(0.5)^2 = 0.196\ \text{m}^2$

$$Q = A \times v$$

$$v = \frac{Q}{A} = \frac{2}{0.196} = 10.204\ \text{m/sec}$$

Head loss due to friction is given by

$$h_f = \frac{fLV^2}{2gD}$$

$$h_f = \frac{0.01 \times 200 \times 10.204^2}{2 \times 9.81 \times 0.5}$$

$h_f = 21.227$ m

Dynamic head is given by

$$H = h_f + H_s$$

$$H = 21.227 + 25$$

$H = 46.227$ m

Combined efficiency of motor and pump is given by

$$0.8 = \frac{gQH}{\text{Total pump capacity required}}$$

Total pump capacity required

$$= \frac{9.81 \times 0.2 \times 46.227}{0.8}$$

$$= 113.37 \text{ kW}$$

Each pump capacity = 22.5 kW

Number of pumps required

$$= \frac{\textit{Total pump capacit required}}{\textit{Each pump capacity}}$$

$$= \frac{113.72}{22.5}$$

= 5.054

= 5 (say). **Ans.**

EXAMPLE 4.14

4.11 POSITIVE DISPLACEMENT PUMPS

4.11.1 Reciprocating Pump

A reciprocating pump is one of the positive displacement pumps which consists of a piston with a cylinder. The piston connected to a crank through connecting rod and the crank is driven by driver where a rotary motion is converted into reciprocating motion of the piston inside the cylinder as shown in Fig. 4.19. The sump is connected to the cylinder through a suction pipe and suction valve. At the exit of the cylinder delivery pipe is connected with the cylinder through a delivery valve.

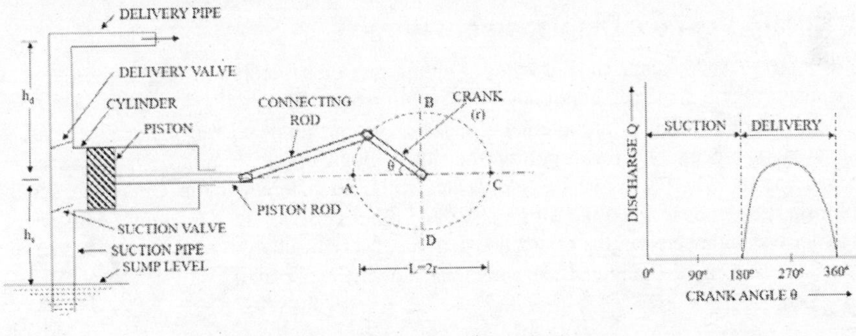

Fig. 4.19: Single acting reciprocating pump

During the operation of reciprocating pump, piston moves backward draws the water inside the cylinder through suction valve while delivery valve remains closed. During the forward stroke of piston, the fluid inside the cylinder gets pressurized and the delivers inside the deliver pipe while suction valve remains closed. Reciprocating pumps can be single acting or double acting. The crank turns from 0° to 180° degrees during the suction stroke, while the piston moves from the far left to the far right. The piston travels inward from its extreme right position toward left when the crank turns from 180° to 360°. When the liquid pressure inside the cylinder exceeds air pressure owing to the piston's inward movement while the suction valve closes and the delivery valve opens. The liquid is gradually increased to the desired height by being pumped up the delivery pipe. Since the liquid is actually delivered to the desired height during this pump action, it is known as its delivery stroke. The crank is at = 0° or 360° (i.e., at its inner dead centre) at the completion of the delivery stroke, the piston is in the extreme left position, one full rotation has been completed while both the suction and delivery valves are closed. As the crank rotates, the same cycle is repeated. Liquid enters a double acting pump from both sides of the piston is shown in Fig. 4. 20.

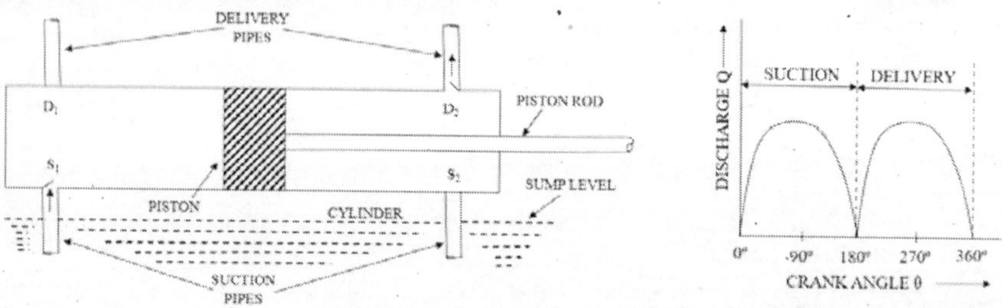

Fig. 4.20: Double acting reciprocating pump

The delivery valve at the front of the piston opens as it moves and pumping liquid into the delivery pipe. Liquid enters the rear side of the cylinder at the same time that suction valve on the back opens. As the piston moves backward the liquid is pushed into the delivery pipe, and at same time the suction valve is opened. The capability to deliver the liquid in a more consistent flow comes with the double acting pumps.

4.11.2 Rotary Positive Displacement Pump

In case of rotary positive displacement pump, rotating motion of gears creates the positive head and simultaneously fluid is displaced continuously in one direction. This is also known as Gear pump and other pumps under this category are internal gear pump, screw pump and vane pump. These are used in hydraulic systems especially for transporting the slurry, clogs, oil etc. Fig. 4.21 shows a schematic of external gear pump. Where two gears are connected with two different shafts. Out of these two gears, one of the shafts is connected to the driver shaft and other acts as idle. The liquid is displaced from suction to delivery side between the casing and gear teeth. The discharge can be calculated considering the volume displaced by the pump (ALn) using the following expression

$$Q = \frac{A\,L\,n\,N}{60} \tag{4.40}$$

Where A is the area between two adjacent teeth, L is the axial length of teeth, n is number of teeth in each gear, and N is r.p.m.

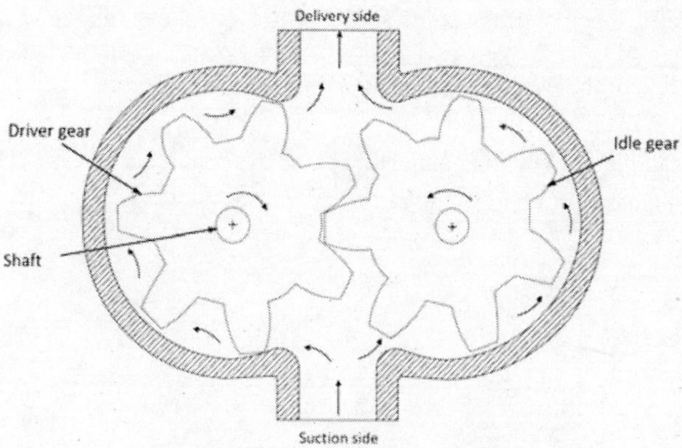

SCAN ME
For better understanding positive displacement pump

Fig. 4.21: Schematic of gear pump

UNIT SUMMARY

- ***Centrifugal pump***

  Centrifugal pumps are rotodynamic machine that generate dynamic pressure to allow liquids to be lifted from lower level to higher level.

  It suitable for variable head and variable discharge.

- ***Work done by the pump***

  The head imparted by the impeller to the liquid is given as

  $$= \frac{1}{g} V_{u_2} u_2$$

  Where,
  V_{u_2} is velocity of whirl at outlet and
  u_2 is tangential velocity of impeller at outlet
  A centrifugal pump's head can be expressed as
  (i) Static head
  (ii) Manometric head

  Static head (H_s)

  H_s is the vertical distance between the liquid surfaces in the sump being raised and the tank to which the liquid is being delivered by the pump. Thus

  $$H_s = h_s + h_d$$

where h_s is the vertical height of the pump shaft's centre line above the liquid surface in the sump and h_d is the vertical height of the liquid surface in the tank above the centre line of the pump shaft to which the liquid is delivered.

Manometric head (H_m)

H_m is the total head required by the pump to meet the external requirements. Thus

$H_m = \frac{1}{g} V_{u_2} u_2 -$ (losses of head in the pump)

- ***Head and Efficiencies***

The efficiencies of a centrifugal pump are

(i) Hydraulic efficiency

$$\eta_{hydraulic} = \frac{\rho g Q H}{P_{impeller}}$$

(ii) Volumetric efficiency

$$\eta_{volumetric} = \frac{Q_{total} - Q_{leakage}}{Q_{total}}$$

(iii) Mechanical efficiency

$$\eta_{mechanical} = \frac{P_{impeller}}{P_{shaft}}$$

(iv) Overall efficiency

$$\eta_0 = (\eta_{\text{hydraulic}}) \times (\eta_{\text{volumetric}}) \times (\eta_{\text{mechanical}})$$

- ***Specific speed***

Specific speed (N_s) of a pump can be defined as the speed of pump to deliver unit discharge against a unit head.

The specific speed is determined by

$$N_s = \frac{N\sqrt{Q}}{H_m^{3/4}}$$

Based on specific speed, pumps are classified as;

(v) Radial flow pump (N_s = 10 to 40)

(vi) Mixed flow pumps with outlet edge parallel to machine axis (N_s= 40 to 80)

(vii)Mixed flow with outlet edge inclined to machine axis provided with volute chamber (N_s = 80 to 100)

(viii) High specific speed axial flow pumps delivering axially, provided with vanes (N_s = 100 to 1000)

The following conditions must be met for complete similarity between the model and the actual centrifugal pump (prototype):

(i) $\left(\frac{N\sqrt{Q}}{H_m^{3/4}}\right)_m = \left(\frac{N\sqrt{Q}}{H_m^{3/4}}\right)_p$

(ii) $\left(\frac{\sqrt{H_m}}{DN}\right)_m = \left(\frac{\sqrt{H_m}}{DN}\right)_p$

(iii) $\left(\frac{Q}{D^3N}\right)_m = \left(\frac{Q}{D^3N}\right)_p$

(iv) $\left(\frac{P}{D^5N^3}\right)_m = \left(\frac{P}{D^5N^3}\right)_p$

- ***Cavitation***

  Thoma's cavitation factor (σ), as used in turbines as well as pumps, is used to predict whether cavitation will occur. To prevent cavitation, the value of σ should not be less than the critical value σ_c given by

  $$\sigma_c = 0.103\left(\frac{N_s}{1000}\right)^{\frac{4}{3}}$$

  where

  N_s = specific speed of the pump.

- ***Performance curves for centrifugal pump***

  For centrifugal pumps, the following three types of characteristics curves are presented in order to predict the behaviour and performance of the pump under varying conditions.

  (i) Main characteristics curves
  (ii) Operating characteristics curves
  (iii) Constant efficiency curves

- ***Reciprocating pump***

  Reciprocating pump is suitable for high head and low discharge which consists of a cylinder with a piston, a suction pipe, a delivery pipe, a suction valve and a delivery valve.

EXERCISES

Multiple Choice Questions

4.1. If D is the impeller diameter of a centrifugal pump, the Discharge through the pump is proportional to ....

(a) D
(b) D^2
(c) D^3
(d) $D \times 1/D^3$

4.2. In a centrifugal pump liquid enters over the impeller

(a)At the top (b) At the bottom

(c) At the centre (d) From sides

4.3. Pump is recommended for low discharge at high head

(a) Radial flow (b) Axial flow

(c) Mixed flow (d) Reciprocating

4.4. Low specific speed pumps are

(a) Axial flow (b) Radial flow

(c) Mixed flow (d) Reciprocating

(e) All of the above

4.5. Suction head for a centrifugal pump is decided by

(a) Delivery head (b) Discharge

(c) Speed (d) Net positive suction head

4.6. The maximum efficiency of a centrifugal pump could be

(a) 30% (b) 50%

(c) 85% (d) 100%

4.7. Centrifugal pumps mostly designed for

(a) Variable head (b) Variable discharge

(c) Constant head and constant discharge (d) Variable head and variable discharge

4.8. Cavitation factor is expressed by

(a) η (b) β

(c) σ (d) ω

4.9. In reverse mode, a centrifugal may act as

(a) Pelton turbine (b) Reciprocating pump

(c) Reaction turbine (d) None of the above

4.10. The specific speed of a centrifugal pump is expressed by

(a) $\frac{N\sqrt{Q}}{H^{1/4}}$ (b) $\frac{N\sqrt{Q}}{H^{5/4}}$

(c) $\frac{P\sqrt{Q}}{H^{3/4}}$ (d) None of the above

4.11. In comparison to a centrifugal pump, reciprocating pumps are used for

(a) High discharge high head (b) Low discharge low head

(c) Low discharge high head (d) High discharge low head

4.12. Reciprocating pump is a ....

(a) Negative displacement pump (b) Positive displacement pump

(c) Diaphragm pump (d) Emulsion pump

4.13. Cavitation is a phenomenon that can take place in

(a) Pelton turbine (b) Tangential flow turbine

(c) Reciprocating pump (d) Centrifugal pump

4.14. Power increases by doubling the pump's speed

(a) 2 times (b) 4 times

(c) 8 times (d) 6 times

4.15. The flow in a centrifugal pump's volute casing outside the rotating impeller is

(a) Axial flow (b) Free vortex flow

(c) Forced vortex flow (d) Radial flow

Answers of Multiple-Choice Questions

4.1 (b), 4.2 (c), 4.3 (d), 4.4 (c), 4.5 (d), 4.6 (c), 4.7 (d), 4.8 (c), 4.9 (c), 4.10 (c), 4.11 (c),4.12 (b), 4.13 (d), 4.14 (c), 4.15 (b).

Short and Long Answer Type Questions

4.1. What is the working principle of a centrifugal pump?

4.2. Define and classify the centrifugal pumps

4.3. How the centrifugal pumps are specified?

4.4. Draw a schematic of a typical centrifugal pump to show its main components

4.5. What are the functions of a casing and impeller of a centrifugal pump?

4.6. Discuss various types of impellers

4.7. Define Net Positive Suction Head (NPSH)

4.8. Discuss the cavitation in centrifugal pumps

4.9. For sizing a centrifugal pump which design parameters are considered

4.10. Define section head, frictional head losses and dynamic head

4.11. Draw the operating characteristics curve for a centrifugal pump

4.12. Define the specific speed.

4.13. On what basis pumps are selected?

4.14. Which pump will be suitable for a unit head and unit discharge and speed of 120 r.p.m?

4.15. Define the conversion factors used to select a pump as turbine

4.16. Why a reciprocating is called as positive displacement pump?

4.17. Define the cavitation factor for a pump.

4.18. What is similarity law of pumps?

Numerical Problems

4.19. Find out the work done by the impeller with the help of velocity triangles for unit discharge, if the water enters the impeller radially with constant flow velocity. Consider impeller internal and external diameters as 100 mm and 300 mm respectively with vane tip angles at inlet and outlet respectively as 25° and 35°. Assume, the pump speed as 1000 r.p.m.

4.20. A pump is installed having the following parameters as: (i) impeller diameter: 300 mm (ii) speed: 1000 r.p.m (iii) head: 10 m (iv) vane angle: 400 and (v) manometric efficiency: 78%.

4.21. Find out flow velocity at outlet, water velocity leaving the vane, angle of absolute velocity at outlet and discharge.

4.22. A pump requires a power of 58 kW to pump the water discharge of 200 lps against a head of 25 m. Determine the overall efficiency of the pump.

4.23. Determine the inlet tip angle for centrifugal pump impeller whose outer diameter is 1.5 times of inner diameter and speed is 1000 r.p.m. Consider the total head of pump as 40 m and constant flow velocity over Impeller as 2 m/s. Other dimensions are as outer diameter as 500 mm, outlet width of 40 mm and vane outlet tip angle as 40°. Also find out the manometric efficiency and work done by the impeller for unit discharge.

4.24. Find out the discharge delivered by a pump at 1000 rpm if the pump discharge at 600 rpm is 40 lps. If diameter of a centrifugal pump impeller was designed to deliver a discharge of 120 lps. Determine the impeller diameter for a discharge of 200 lps.

4.25. A centrifugal pump is required to deliver water for a flow rate of 150 lps from a sump to a water overhead tank which is kept at a height of 25 m from the sump. Calculate the power required to operate the pump if friction losses in the pipes are considered as negligible.

4.26. If power required for a centrifugal pump with 0.2 m impeller diameter is 18 kW then how much power will be required if the impeller diameter is reduced by 8%.

4.27. For a water pumping system, one pump was installed to deliver a discharge of 100 lps against a head of 50 m. Assume the efficiency of the motor required to drive the pump is 85%. After expansion of the pumping system, the required quantity of water is 400 lps. Suggest the number of pumps and their operation in series or parallel. Determine the power required, if the efficiency of each pump is 80%.

4.28. For a water pumping system for pillar furnace, one pump, having specification; discharge, Q =12.5 lps, head, H= 30 m and power, P= 6.7kW was selected. Determine the efficiency of the pump. As per the pillar furnace manufacturer water required for the cooling is 12.5 lps, but at 60 m. How many pumps are required for the requirement? How much energy can be saved if a new pump having 65% efficiency for the new requirement?

4.29. A Centrifugal pump is installed to supply water from a reservoir to a tank which is at a vertical height of 12 m above it. The diameter of the suction pipe is 300 mm and it is 6 m long. At the end of the suction pipe which extends under the water level in the reservoir, are fitted a foot valve (loss coefficient, $k_v = 1.5$). There is one 90° bend, (loss coefficient, kv=0.8) on the side of suction. The discharge pipe is 250 mm diameter and 135 m long. It is fitted with a gate valve and two medium radius 90° bends (loss coefficient k_v=0.9). Find the power of the pump to discharge 85 lit/sec of water considering all losses of head in pipe fittings. The overall efficiency of pump is 70%. Find also the capacity of driving motor. Take friction factor for pipe as 0.014.

4.30. Determine the head losses and draw a system resistance curve along with the operating curve for the pump for the pumping system through which the velocity is 3 m/s and consisting of the followings; (i) foot valve to suction pipe (loss coefficient, kv = 1.85) (ii) suction pipe (length = 4m; dia = 100 mm, friction factor f = 0.038) (iii) bend in the suction pipe (kv = 1.90) (iv) centrifugal pump delivery pipe (length = 5m; dia = 100 mm, friction factor f = 0.038) before a 90° bend along with a control valve (k_v = 0.19) 90° bend (kv = 1.90) and (v) delivery pipe (length = 1m; dia = 100 mm, friction factor f = 0.038). after the bend along with a control valve (kv = 0.19). Also find out the power required by the pump having efficiency as 75% to deliver a discharge of 23.5 lps.Which pump will be suitable to lift the water discharge of 0.035 $\frac{m^3}{sec}$ against a total head of 25 m. having a speed of 300 r.p.m?

4.31. Determine the efficiency of a pump which is driven by an electrical motor of 7.5 kW and deliver a discharge of 35 lps against a head 35 m.

4.32. Determine the specific speed of a pump for a given discharge of 0.1 m^3/sec and head of 25 m if the speed of the pump 1000 r.p.m. Find out the discharge, input power and head of the pump at rpm of 1500, assuming the efficiency of pump as 85%.

4.33. A pump having an impeller diameter of 0.30 m, discharge capacity of 50 lps and head of 25 m. Find out head and impeller diameter of a similar pump for a discharge of 25 lps.

4.34. A model of centrifugal pump having runner diameter of 20 cm and running width 1000 r.p.m to deliver a discharge of 150 lps against a head of 4.8m at its best efficiency point and it requires 7.5 kW power.

4.35. Based on this model find out (i) Impeller diameter (ii) speed of the prototype to deliver 9.5 cm3/sec discharge against a head of 5.5 m.

4.36. Determine the efficiency of a three-stage centrifugal pump which is running at 600 r.p.m. to develop a total head of 40 m to discharge water of 0.10 m^3/s. If the width and diameter of each impeller at outlet is 5 cm and 60 cm respectively having vane tip angle of 40°. Find the manometric efficiency of pump.

4.37. Calculate the net power suction head (NPSH) for a centrifugal pump having speed of 1000 r.p.m. delivers 0.20 cumec. of water against a head of 30m. Consider the frictional losses in the pipe as 0.2 m. Assume the suitable values for other parameters.

4.38. A model having scale of 1/10 is constructed to determine the best design of a pump to deliver 50 m3/s discharge against a net head of 9 m when running at 100 rpm. If head available at laboratory is 6 m and the efficiency is 85% find (i) running speed of model (ii) flow required in laboratory (iii) power required for the model (iv) specific speed in each case.

4.39. Find out the cavitation factor for a pump installed at a location having suction head as 5 m, head losses due to friction as 0.5 m and manometric head as 25. Assume the suitable values for other required parameters.

4.40. In order to avoid the cavitation, find out the value of critical Thoma's coefficient for a pump required to deliver a discharge of 500 lps against a total head of 5 m. Take the speed of pump as 1500 r.p.m.

Applications

The applications of pumps are very wide which include agriculture, industries, and residential water supply. The unit-4 will be beneficial for the students to understand the basics of designs, constructions, operation and selection of pumps for a given requirement. Further, this unit will beneficial for optimal selection of centrifugal pumps to the industries and agriculture applications.

REFERENCES AND SUGGESTED READINGS

List of some of the books is given below which may be used for further learning of the subject:

1. E. Mosonyi, Water Power Development, Vol. I and II, Nem Chand and Brothers, 2009.
2. P.S. Nigam, Handbook of Hydroelectric Engineering, Nem Chand and Brothers, 2001.
3. J. Lal, Hydraulic Machines, 3rd edition (reprint), Metropolitan Book Co. Private Limited, 2002.
4. National and International Standards
5. G. Brown, Hydro-electric Engineering Practise, Vol. II, CBS Publication, 1984.
6. Yunus A. Çengel, Fluid Mechanics: Fundamentals and Applications, McGRAW-HILL publication, 2006.
7. P. N. Modi, Hydraulics & Fluid Mechanics including Hydraulics Machines, Rajsons Publications Pvt. Ltd., 2014.

5 Fluid Machines – Water Turbines

UNIT SPECIFICS

Through this unit we have discussed the following aspects:

- *Working principle of water turbine and their classifications;*
- *Components of different water turbines;*
- *Different terms and velocity triangles;*
- *Performance characteristics of water turbines;*
- *Governing of water turbines;*
- *Similarity law and specific speed;*
- *Selections of water turbines for different site conditions;*

The practical applications of the topics are discussed for generating further curiosity and creativity as well as improving problem solving capacity.

Besides giving a large number of multiple choice questions as well as questions of short and long answer types marked in two categories following lower and higher order of Bloom's taxonomy, assignments through a number of numerical problems, a list of references and suggested readings are given in the unit so that one can go through them for practice. It is important to note that for getting more information on various topics of interest some QR codes have been provided in different sections which can be scanned for relevant supportive knowledge.

After the related practical, based on the content, there is a "Know More" section. This section has been carefully designed so that the supplementary information provided in this part becomes beneficial for the users of the book. This section mainly highlights the initial activity, examples of some interesting facts, analogy, history of the development of the subject focusing the salient observations and finding, timelines starting from the development of the concerned topics up to the recent time, applications of the subject matter for our day-to-day real life or/and industrial applications on variety of aspects, case study related to environmental, sustainability, social and ethical issues whichever applicable, and finally inquisitiveness and curiosity topics of the unit.

RATIONALE

This fundamental unit on water turbines helps students to get a primary idea about the working principle, characteristics and selections of water turbines for field application of these under different operating conditions. It explains the impulse- momentum equation to know the working principle of water turbines. It then explains the various components of different water turbine and their velocity triangles to understand the procedure to determine the output of water turbines. To understand the behaviour of water turbine under given conditions, characteristics curves are discussed. Governing of water turbines is included in the unit. Specific speed and similarity law are discussed which will help in selection of water turbines. Stepwise selection of water turbines is also given

Water turbines is an important unit under fluid machines essentially deals power generation from water energy available in the form of potential and kinetic energy. This permits to understand the working principle, construction, characteristics and selection of water turbines to help the students in studying these machines under advanced courses and with industries.

PRE-REQUISITES

Science: Properties of fluids especially water such as density, specific gravity, viscosity, flow velocity etc. (Class XII)

Physics: Fundamental of energy conversion process such as conservation of mass (continuity equation) and energy (Bernoulli's equation) (Class XII)

Mathematics: Vector analysis such as velocity and force diagrams (Class XII)

UNIT OUTCOMES

List of outcomes of this unit is as follows:

U5-O1: *Describe the working principle and classification of water turbines*

U5-O2: *Draw the velocity triangles and evaluate the performance of Pelton, Francis and Kaplan turbines*

U5-O3: *Define the function of draft tube and governors*

U5-O4: *Describe the performance characteristics of various water turbines*

U5-O5: *Describe the selection procedure of water turbines and solve water turbines related problems*

Unit-5 Outcomes	EXPECTED MAPPING WITH COURSE OUTCOMES (1- Weak Correlation; 2- Medium correlation; 3- Strong Correlation)					
	CO-1	CO-2	CO-3	CO-4	CO-5	CO-6
U5-O1	3	3	2	-	3	1
U5-O2	2	1	-	-	2	2
U5-O3	2	-	3	2	2	1
U5-O4	-	2	2	3	1	-
U5-O5	3	1	3	2	-	2

5.1 WATER TURBINE

To extract the available energy of water for power generation a prime mover is required which is a device to convert one form of energy into another form of energy. Here, water (hydro) turbine is the prime mover used to convert the potential energy available in water into mechanical power. Further, the mechanical power developed by the turbine is then used to derive the alternators to generate electricity in a hydro power station.

5.2 CLASSIFICATION OF WATER TURBINES

Energy in water is basically available in the form of potential energy, which is estimated from the basic measured parameters: head (height) and discharge (volume per unit time) of water. According to quantity of these parameters, water turbines are categorised based on the following criterion;

A. Based on head and quantity of water (discharge)
   (i) Impulse turbine
      Pelton turbine: High head and low discharge
   (ii) Reaction turbine
      a. Francis turbine: Medium head and medium discharge
      b. Kaplan turbine: Low head and high discharge

SCAN ME
For better understanding water turbine classification

B. Based on action of water over the runner
   (i) Impulse turbine (Pelton turbine)
   (ii) Reaction turbine
      a. Francis turbine
      b. Axial flow turbine (Propeller and Kaplan)

Water turbine classification based on action of water over the runner as shown in Fig. 5.1.

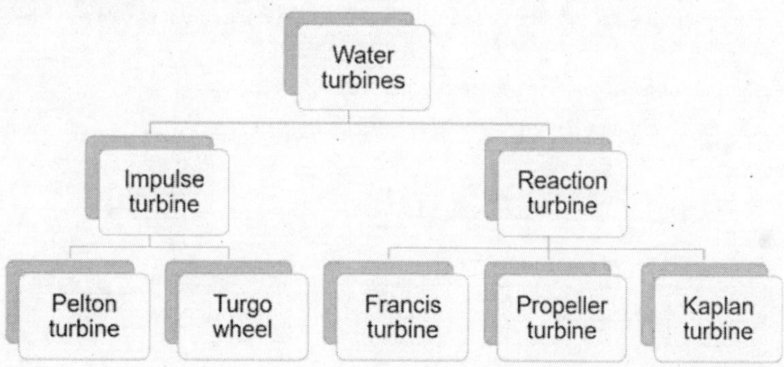

Fig. 5.1: Classification of water turbine according to action of water moving over blade

C. Based on direction of flow of water over the runner
   (i) Tangential flow turbine (Pelton turbine)
   (ii) Radial flow turbine (old Francis turbine)

(iii) Mixed flow turbine (Modern Francis turbine)
(iv) Axial flow turbine (Kaplan turbine)

Water turbines classification based on direction of flow of water over the runner is also given in Fig. 5.2.

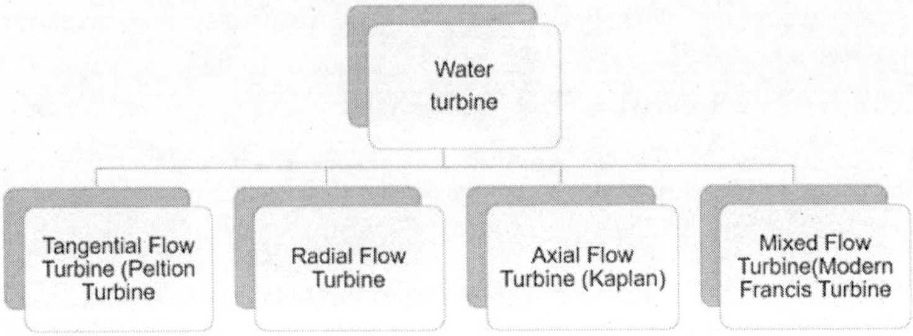

Fig. 5.2: Classification of water turbine according to water flow direction in the runner

D. Based on the shaft orientation
   (i) Vertical (Pelton, Francis and Kaplan)
   (ii) Horizontal (Pelton, Francis and Kaplan)
E. Based on specific speed (N_s)
   (i) N_s = 12-35 (Single jet Pelton turbine)
   (ii) N_s = 80-400 (Francis turbine)
   (iii) N_s = 340-1000 (Kaplan turbine)

5.3 WORKING PRINCIPLE OF WATER TURBINES

The water turbine works based on impulse momentum equation which includes two important terms, (i) static pressure and (ii) dynamic force. The static pressure implies no fluid motion whereas dynamic force involves always a change of momentum. Basically, Newton's second law of motion used for impulse-momentum equation derivation which states "The rate of change of momentum is proportional to the applied force and takes place in the direction of force".

The momentum of a body is the product of its mass and velocity.

If m be the mass of water having velocity v with change in velocity of dv over a time dt.

It is therefore,

$$\text{Change in momentum} = m \, . \, dv \tag{5.1}$$

$$\text{Rate of change in momentum} = m \, . \frac{dv}{dt} \tag{5.2}$$

According to Newton's Second Law

Dynamic force = rate of change in momentum

$$F = m \, . \frac{dv}{dt} \tag{5.3}$$

$$F . dt = m . dv \tag{5.4}$$

Thus, it is shown that the impulsive force is equal to total momentum change of the body.

Some practical examples based on impulse- momentum equation can be discussed as follows:

Case I: Force exerted by water jet on a stationary flat plate

Consider the water jet from a nozzle with a velocity (v) strikes over a flat plate which is kept stationary and perpendicular to the water jet as shown in Fig. 5.3.

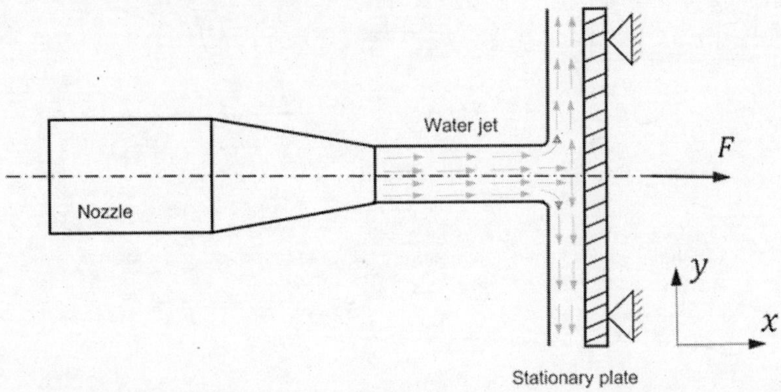

Fig. 5.3: Schematic diagram of force exerted by water jet on a stationary flat plate

Let Q be the quantity of water falling on the plate in m^3/s and mass, m of water is expressed in Eq. (5.5) as

$$m = \rho Q \tag{5.5}$$

Where, ρ is the density of water.

If the fluid is so deflected by the plate that it loses all the velocity v normal to the plate, then

Dynamic force = Change of momentum/sec

= (Mass striking the plate/sec) x (Change of velocity normal to the plate)

$$F = \rho Q \,.\, (v - 0)$$

$$F = \rho Q \,.\, v \tag{5.6}$$

Equation of continuity is written as,

$$Q = a \,.\, v \tag{5.7}$$

Where, a is the cross-sectional area of the jet.

Then the dynamic force exerted by water jet on a stationary flat plate can be written as,

$$F = \rho\, a\, v^2 \tag{5.8}$$

Case II: Force exerted on inclined stationary plate

Consider the water jet from a nozzle with a velocity (v) strikes over a inclined plate which is kept stationary as shown in Fig. 5.4.

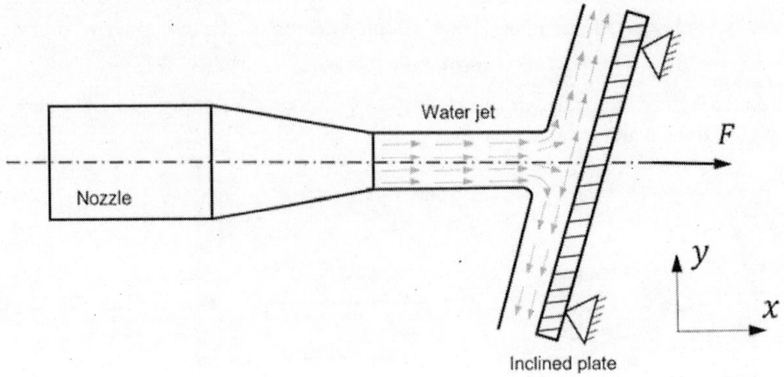

Fig. 5.4: Schematic diagram of force exerted by water jet on inclined stationary plate

$$F = \rho\, Q\, v \sin\theta \tag{5.9}$$

$$F = \rho\, a\, v^2 \sin\theta \tag{5.10}$$

Component of this force F in the direction of jet is

$$F_x = F\, sin\theta = \rho\, a\, v^2 \sin^2\theta \tag{5.11}$$

No work is done in these cases as plate is not moving.

EXAMPLE 5.1

Example 5.1 A Water jet of diameter 45 mm strikes a stationary plate in such a way that the angle between the plate and the jet is 25°. The force exerted in the direction of the jets is 1500 N. Determine the discharge of water.

Solution

Given data:

Diameter of the jet, $d = 0.045$ m

Area, $a = \frac{\pi}{4}(0.045)^2 = 0.00159 \text{ m}^2$

Angle, $\theta = 15^0$

Force in the direction of jet, $F_x = 1500$ N

We know that the force in the direction of jet is given by the equation as

$$F_x = \rho a v^2 \sin^2\theta$$

$$1500 = 1000 \times 0.00159 \times v^2 \times sin^2 25°$$

$$= 0.1065\, v^2$$

$$v = 72.677 \text{ m/s}$$

Discharge is given by

$$Q = \text{area} \times \text{velocity}$$
$$= 0.00159 \times 72.677$$
$$= 0.1155\ m^3/s$$
$$= 115.5\ \text{litres/s.}$$ **Ans.**

EXAMPLE 5.1

Case III: Force exerted on moving flat plate

Consider the water jet from a nozzle with a velocity v strikes over a moving flat plate with velocity u as shown in Fig. 5.5.

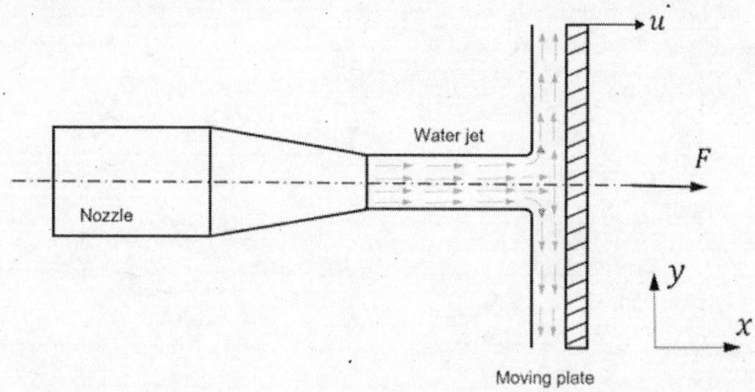

Fig. 5.5: Schematic diagram of force exerted by water jet on moving flat plate

Let the plate moves with a velocity u in the same direction as the jet. As the water jet will come in contact with the plate, it will attain the velocity of plate u in the same direction and the change in velocity will be as $(v - u)$. The corresponding quantity of water will be as;

i.e. $$Q = a\,(v - u) \tag{5.12}$$

Therefore, force will be expressed by the Eq. (5.13) as

$$F = \rho\, Q(v - u) \tag{5.13}$$

$$F = \rho\, a\,(v - u)(v - u)$$

$$F = \rho\, a\,(v - u)^2 \tag{5.14}$$

The work done by the plate will be expressed in the following Eq. (5.15) as

$$F = \rho.\ a(v - u)^2.\text{u} \tag{5.15}$$

EXAMPLE 5.2

Example 5.2 If a jet of water having diameter of 5 cm and strikes over a flat plate at 90^0 with a velocity of 10 m/sec. The plate is moving with a velocity of 3 m/sec, Find out

(a) Force acting over the plate

(b) Work done by the plate

Solution

Given data:

Jet diameter, $d = 5$ cm

Water velocity, $v = 10$ m/sec

Velocity of plate, $u = 3$ m/sec

(a) Force acting over the plate

As per the Impulse-momentum equation

$$\begin{aligned} \text{Force,} \quad F &= \text{rate of change of momentum} \\ &= \text{mass} \times \text{change in velocity} \\ &= \rho A(v-u)((v-u)-0) \\ &= 1000 \times \frac{\pi}{4} \times (0.05)^2 \times (10\text{-}3)^2 \\ &= 785.4 \text{ N. } \textbf{Ans.} \end{aligned}$$

(b) Work done by the plate

$$\begin{aligned} \text{Work done,} \quad W &= \text{Force} \times \text{velocity of plate} \\ &= F \times u \\ &= 785.4 \times 3 \\ &= 2356.2 \text{ W. } \textbf{Ans.} \end{aligned}$$

Case IV: Water jet strike over a series of flat plate mounting over the periphery of a wheel

As continuous work cannot be performed by the system discussed in case III, it is therefore practical system can be obtained by considering a series of plates mounted over a wheel as shown in Fig. 5.6.

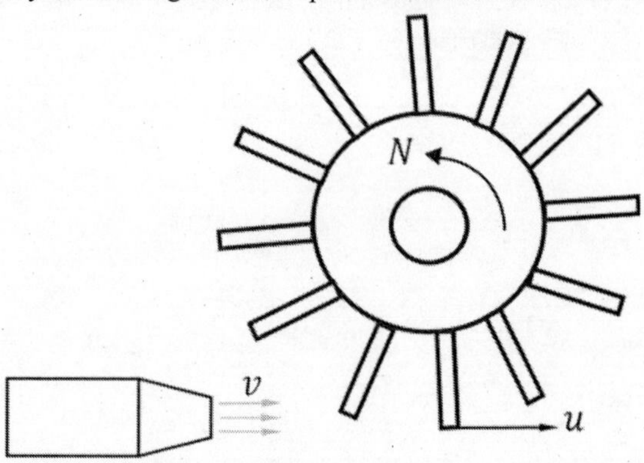

Fig. 5.6: Water jet strike over a series of flat plate mounting over the periphery of a wheel

Based on the concept discussed above cases, the continuously work done will be as expressed by Eq. 5.16 as;

$$W = F\,.u = \rho\, a\, v\, (v-u)u \tag{5.16}$$

The efficiency of such a system can be discussed as

The input in the form of kinetic energy of the jet $= \frac{1}{2}\rho a v v^2$

The efficiency of the system will be based on work done by the system and kinetic energy. It can be expressed as

$$\eta = \frac{2\ \rho\ a\ v(v - u)u}{\rho\ a\ v\ v^2} \tag{5.17}$$

$$\eta = \frac{2\ (v - u)u}{v^2} \tag{5.18}$$

For maximum efficiency η_{max}, $\frac{d\eta}{du} = 0$

$$\frac{d}{du}\left(\frac{2\ (v-u)\ u}{v^2}\right) = v - 2u \tag{5.19}$$

$$u = v/2$$

Substituting this value in Eq. (5.18), we get

$$\eta_{max} = \frac{2\ \left(v - \frac{v}{2}\right) \cdot \frac{v}{2}}{v^2} \tag{5.20}$$

$$\eta_{max} = \frac{2\ \frac{v^2}{2} - \frac{v^2}{2}}{v^2}$$

$$\eta_{max} = 1 - \frac{1}{2}$$

$$\eta_{max} = 0.5 = 50\% \tag{5.21}$$

From the above discussions, it is found that maximum efficiency of such a system can be achieved upto 50%.

Case V: Water jet strike over curved stationary plate

The water jet strikes over a curved plate at an angle of α_1 and it deviated to an angle α_2, both angles being measured with respect to x-axis direction (Fig. 5.7). Let v_1 and $v_2$be the velocity of jet at inlet and outlet respectively. The velocity v_1 will be equal to v_2 as long as there is no friction on the plate.

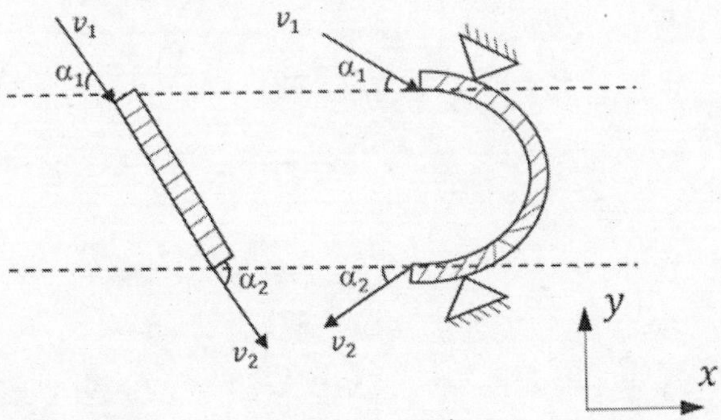

Fig. 5.7: Schematic diagram of Water jet strike over curved stationary plate

Velocity of jet at inlet in x-direction = $v_1 \cos \alpha_1$
Velocity of jet at outlet in x-direction = $v_2 \cos \alpha_2$
Force exerted by the jet on the plate in x-direction can be determined by applying impulse-momentum equation

$$\left(\sum F_x\right) t = m \sum v_x \tag{5.22}$$

Or $$F_x = \frac{m}{t} \text{ (change in velocity in x – direction)}$$

$$F_x = \rho Q\,(v_1 \cos \alpha_1 - v_2 \cos \alpha_2) \tag{5.23}$$

Where, $$Q = a v_1$$

$\therefore$ $$F_x = \rho a\, v_1\,(v_1 \cos \alpha_1 - v_2 \cos \alpha_2) \tag{5.24}$$

Case VI: Water jet strikes over moving curved plate

The force exerted by the water jet over curved plate as discussed above in case V, is simple case to understand. However, it will become complicated in reference to the various velocity components, if the curved plate is moving to get useful work from the system.

In order to get change in momentum over the moving curved plate, resultant change in velocity is to be determined by considering various velocity components. The different velocity components can be managed to get resultant velocity by drawing velocity triangles at inlet and outlet of the curved plate which are discussed as follows;

Various components of velocity at inlet and outlet over a moving curved plate observed are shown in Fig. 5.8.

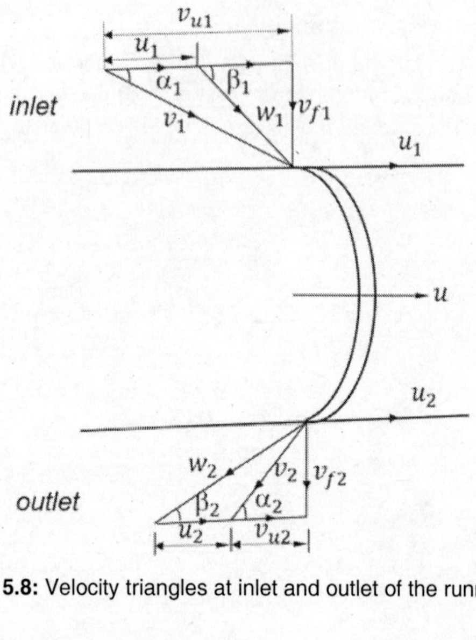

Fig. 5.8: Velocity triangles at inlet and outlet of the runner

Let v_1 and v_2 = Absolute velocities of jet at inlet and outlet, respectively

α_1 and α_2= Angles of jet at inlet and outlets measured from x-direction, respectively

u_1 and u_2= Peripheral/circumferential velocities at inlet and outlet, respectively

w_1 and w_2= Relative velocities at inlet and outlet, respectively

β_1 and β_2 = Angles of the relative velocities in the reversed direction x, respectively

v_{u_1} and v_{u_2} = Velocities of whirl at inlet and outlet, respectively

v_{f_1} and v_{f_2} = Velocities of flow at inlet and outlet, respectively

Force is given by

$$F = \rho Q(v_1 \cos\alpha_1 - v_2 \cos\alpha_2) \quad (5.25)$$

$$F = \rho a(v_1 - u)(v_1 \cos\alpha_1 - v_2 \cos\alpha_2)$$

Work done per unit weight of water striking per sec is given by

$$W = \frac{1}{g}(v_{u1} - v_{u2})u \quad (526)$$

$\rho = 1000 \text{ kg/m}^3$

$g = 9.81 \text{ m/s}^2$

$\therefore$ Efficiency of jet, $\eta = \frac{output}{input}$ (5.27)

$$\eta = \frac{(v_{u1} - v_{u2})\frac{u}{g}}{\frac{v_1^2}{2g}}$$

$$\eta = \frac{2u(v_{u1} - v_{u2})}{v_1^2} \quad (5.28)$$

Example 5.3 A jet of water with a velocity of 30 m/s enters the curved vane in the form of a bucket. The vanes are measured on a wheel having a mean diameter of 1 m. The wheel makes 300 r.p.m. The vane tips make angle of 15^0 and 30^0 to the wheel tangent of inlet and outlet respectively. Find the force exerted by the jet per kg. Water in tangential direction, work done / kg of water and hydraulic efficiency. Assuming that there are no other losses.

EXAMPLE 5.3

EXAMPLE 5.3

Solution

Given data:
Jet velocity, $v = 30$ m/sec
Mean diameter, $D = 1$ m
Speed, $N = 300$ r.p.m

Tangential speed, $u = \frac{\pi D N}{60}$

$$u = \frac{\pi \times 1 \times 300}{60}$$

$$= 15.70 \text{ m/s}$$

$$F_x = \rho Q\,(v_{u_1} - v_{u_2})$$

$$= 29.23 - 3.55 = 25.65 \text{ N/kg}$$

Work done / kg,

$$= F_x \times u$$

$$= 25.65 \times 15.70$$

$$= 402.70 \text{ J/kg. } \textbf{Ans.}$$

Efficiency,

$$\eta = \frac{\frac{\text{workdone}}{\text{kg}} \text{ of water}}{\text{input power}}$$

$$= \frac{402.70}{\frac{30^2}{2}}$$

$$= \frac{402.70}{450}$$

$$\eta = 87.5\%. \textbf{ Ans.}$$

5.4 HEADS AND EFFICIENCIES

5.4.1 Head

The head of a hydroelectric power plant is entirely dependent on natural conditions. Head is characterised as: Gross head and Net or Effective head. Gross head is defined as the difference in water level at the intake, called head race level, and water level at the outlet, called tail race level. Both these levels are measured simultaneously. The water is carried to the inlet of a water turbine through canals and penstock pipes. There are losses when water flows through these canals, tunnels and penstocks, and head losses (friction) are considered to get the net head. Thus, the net head can be expressed by the following the expression as given in Eq. (5.29)

$$H_{\text{net}} = H_{\text{gross}} - h_l \tag{5.29}$$

Where, H_{net} is net head

H_{gross} is gross head

h_l is head loss

SCAN ME

For better understanding head and efficiency

5.4.2 Efficiency

To find out the performance of turbines, the following efficiencies are of significance:

(a) *Hydraulic efficiency*: It is the ratio of developed power by the runner to the power supplied to the runner by the water.

$$\eta_h = \frac{\text{power produced by the runner}}{\rho g Q H} \tag{5.30}$$

Where, Q is the quantity of water flowing through the turbine per unit time, and

H is the net head available at the turbine inlet.

This represents the effectiveness with which energy is transferred from the fluid to the runner.

(b) *Volumetric efficiency*: When water enters the turbine, there is a possibility that some water may go to the downstream without performing the work (i.e., without flowing over the runner blades). Accordingly, the ratio of the volume of water flowing over the runner to the total volume of water supplied to the turbine is termed as volumetric efficiency and expressed as given in Eq. (5.31).

$$\eta_v = \frac{Q - \Delta Q}{Q} \tag{5.31}$$

Where, Q is water flow rate and

ΔQ is water not passing through the runner blades.

(c) *Mechanical efficiency*: The power available at the shaft for use is always less than the power produced by the runner due to the mechanical losses. Mechanical efficiency is thus defined as the ratio of useful power available at the shaft to the power produced by the runner.

$$\eta_m = \frac{\text{power available at the shaft}}{\text{power produced by the runner}} \tag{5.32}$$

(d) *Overall efficiency*: It is defined as the ratio of useful power available at the shaft to the power supplied by water turbine at inlet and is obtained by multiplying the above efficiencies (Eqs. 5. 30 – 5.32) and can be expressed as given in the Eq. 5.33.

$$\eta_o = \frac{\text{power available at the shaft}}{\rho g Q H}$$

$$\eta_o = \eta_h \times \eta_m \times \eta_v \tag{5.33}$$

The hydraulic turbines are directly coupled to electric generators to produce electric power. There is a loss of power in transmission. Coupling and the generators also have some losses. Therefore, the power output at the generator shaft is a little less than the power available at the runner outlet.

EXAMPLE 5.4

Example 5.4 Determine the hydropower plant capacity for the given data of head as 150 m and discharge as 0.5 m^3/sec. Considering overall efficiency as 85%.

Solution

Power equation for hydro power,

$$P = 9.81 \times \text{discharge } (Q) \times \text{head } (H) \times \text{overall efficiency } (\eta_o)$$
$$= 9.81 \times 150 \times 0.5 \times 0.85$$
$$= 625.38 \text{ kW. } \textbf{Ans.}$$

5.5 PELTON TURBINE

The Pelton turbine, is an impulse turbine, which essentially converts pressure energy into kinetic energy in the form of a high-speed water jet coming from a nozzle, that uses hydro(water) potential. The high-speed water jet impacts the runner, imparts most of its energy to the bucket, and then drops into the tail race with minimal remaining energy. This type of turbine runs in the open, and the casing avoids the splashing of water, directs water to the tailrace, and provides safety. The Pelton turbine with its typical components (i) nozzle (ii) runner with buckets, and (iii) casing are shown in Fig. (5.9).

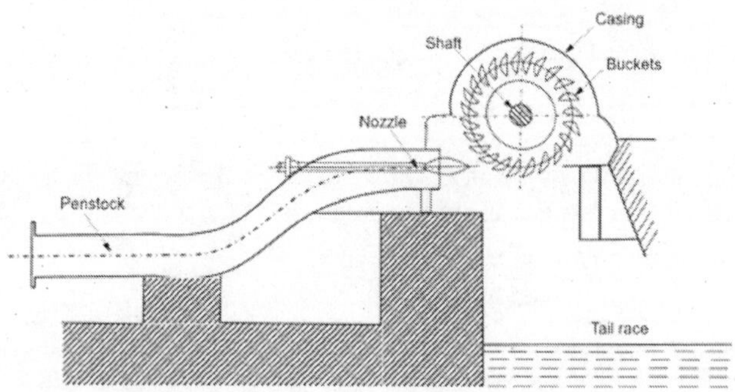

Fig. 5.9: Main components of Pelton turbine

These turbines are more suitable for the sites where very high head and low discharge are available and having the following advantages:

(i) More tolerance of the sediment erosion,
(ii) Easy access to working parts,
(iii) Pressure seals around the shaft not required,
(iv) Easy to fabricate and maintain,
(v) Better part-load efficiency than Francis turbine,
(vi) Power house buildings requires less space.

The main drawback of Pelton turbine is that they are not suitable for low-head sites due to constraints of accommodating larger quantity of water for the given capacity. However, for a given site of more discharge, multi jet Pelton turbine having more nozzles can be used.

5.5.1 Pelton Turbine Components

Followings are the basic components of typical Pelton turbine;

(i) Runner with buckets
(ii) Nozzles
(iii) Casing
(iv) Control valves
(v) Shaft bearing

5.5.1.1 Runner with buckets

The runner (Fig. 5.10) with buckets is provided to convert the kinetic energy of jet into rotational energy of shaft (Mechanical energy). The dimensions of the runner as shown in Fig. 5.11.

Fig. 5.10: Runner of a Pelton turbine

$$v_{jet} = K_v \sqrt{2gH} \tag{5.34}$$

$$u = K_u \sqrt{2gH} \tag{5.35}$$

K_u is speed ratio and K_v is coefficient of velocity.

Theory and experiment show that efficiency is best when the Pitch circle has about half the velocity of jet.

i.e. $$u = \frac{1}{2} v_{jet} \tag{5.36}$$

For designing the runner, K_u is usually taken as 0.46.

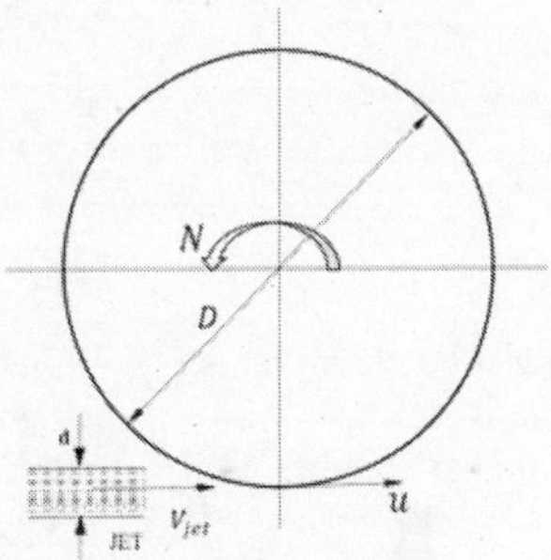

Fig. 5.11: Schematic of Pelton turbine runner

Also $$u = \frac{\pi DN}{60} \tag{5.37}$$

where D = Diameter of runner,

N = revolution per minute

Equating Eq. 5.35 and Eq. 3.37, we get

$$\frac{\pi DN}{60} = K_u \sqrt{2gH}$$

$$D = \frac{38.9 \sqrt{H}}{N} \tag{5.38}$$

Number of buckets may be determined by the Taygun relationship, Fig.5.12 Shown the schematic diagram of Pelton bucket.

$$Z = 0.5\,m + 15 \tag{5.39}$$

Jet ratio, $m = \dfrac{D_{runner}}{d_{jet}}$

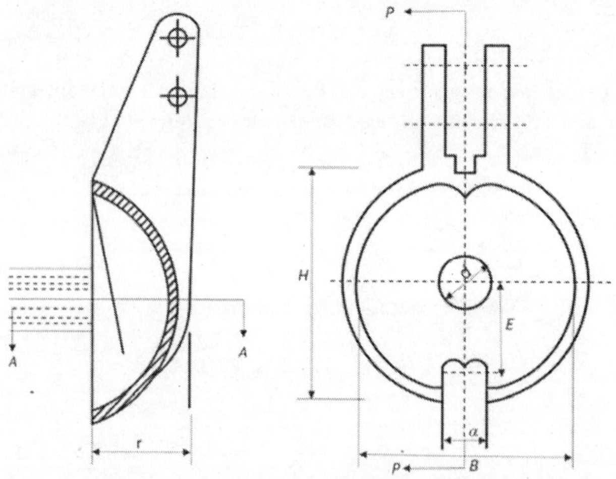

Fig. 5.12: Schematic diagram of Pelton bucket

It must be clearly understood that these dimensions are decided upon certain logical conclusions, but there exists upto now no mathematical method to calculate these dimension. It is obvious that the minimum breadth of bucket required just only to accommodate the incoming and outgoing jet shall be 2d (d is diameter of jet) but the axial distance has also to be provided for the jet to turn and taking this distance as 0.4 - 0.6d on each side.

Therefore, breadth (B) of bucket should be between 2.8 to 3.2 d.

i.e. breadth $\quad B = (2.8 - 3.2)\,d$

It is clear that jet does not always strike the bucket at the same radial distance from the centre of the wheel.

So height of bucket $\quad H = (2.2 - 2.8)\,d$

The function of depth of the bucket, is to provide sufficient tangential distance, so that the jet can turn through desired angle.

Depth $\quad T = (0.6 - 0.9)\,d$

Further, the work done by water and efficiency of the Pelton turbine is explained with the velocity triangle is shown in Fig. 5.13.

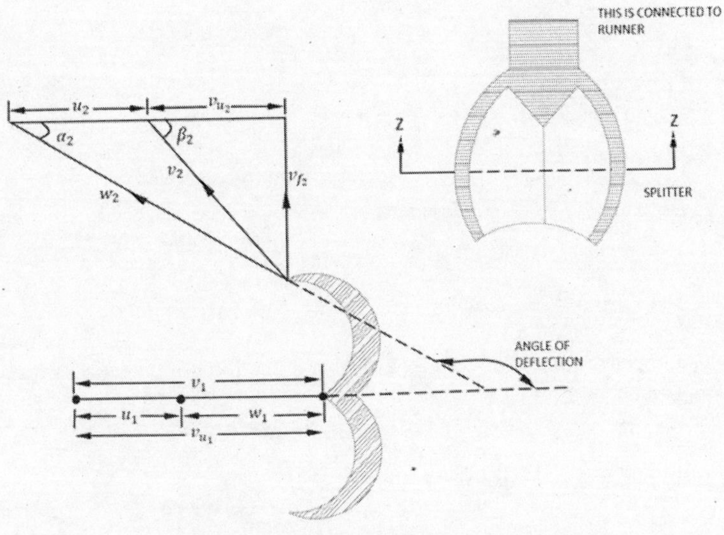

Fig. 5.13: Velocity triangle diagram of Pelton turbine

Force acting on the bucket = Change in Momentum

$$= mdv \quad (5.40)$$

$$= m(v_{u1} \pm v_{u2})$$

$$= m(v_1 + v_2 \cos\beta_2) \quad (5.41)$$

Power given to the runner by the jet is given by

$$= m(v_1 + v_2 \cos\beta_2)\, u \quad (5.42)$$

As $u = u_1 = u_2 = \frac{\pi DN}{60}$

Kinetic energy of the jet $= \frac{1}{2} m v_1^2$ (5.43)

$$= \frac{1}{2}(\rho a v_1) v_1^2$$

$$= \frac{1}{2}\rho a v_1^3 \quad (5.44)$$

Hydraulic efficiency $(\eta_h) = \frac{\text{Power output}}{\text{K.E.of jet per second}}$ (5.45)

$$\eta_h = \frac{m(v_1 + v_2 \cos\beta_2)u}{\frac{1}{2} m v_1^2}$$

$$\eta_h = \frac{2(v_1 - u)[1 + \cos\beta_2]u}{v_1^2} \quad (5.46)$$

For efficiency to be maximum,

$$\frac{d\eta_h}{du} = 0$$

$$\frac{d\eta_h}{du} = 2\left(\frac{1 + \cos\beta_2}{v_1^2}\right)\frac{d}{du}\{v_1 u - u^2\} = 0$$

$$u = v_1/2 \tag{5.47}$$

By putting value of u in Eq. 5.46,

$$\eta_h = \frac{1 + \cos\beta_2}{2} \tag{5.48}$$

5.5.1.2 Nozzle

Nozzle is provided to convert the hydro energy available in the form of potential energy into kinetic energy as shown in Fig. 5.14.

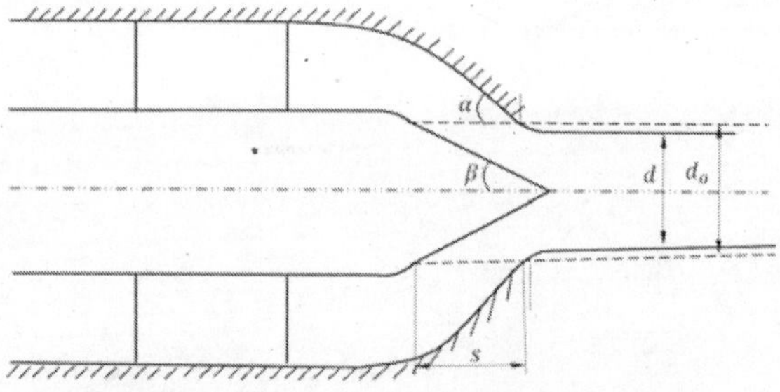

Fig. 5.14: Nozzle with spear valve of Pelton turbine

The quantity of water required is given by,

$$Q = \frac{P}{9.81\, H\, \eta} \tag{5.49}$$

Where,

P is the output power in kW

H is the head in meters of water column

η is the overall efficiency

Total discharge is given by

$$Q = \frac{\pi}{4} d^2 \times v_j \times n_j \tag{5.50}$$

Where,

d is the diameter of jet

v_j is the velocity of jet

n_j is number of jets

Total discharge is calculated by

$$\frac{Q}{n_j} = \frac{\pi}{4} d^2 \times v_j \tag{5.51}$$

$$q = \frac{\pi}{4} d^2 \times v_j \qquad \left(q = \frac{Q}{n_j}\right)$$

Also,

$$v_{jet} = K_v(\sqrt{2gH}) \tag{5.52}$$

K_v is the coefficient of velocity introduced due to the losses in the nozzle.

$$q = \frac{\pi}{4} d_j^2 \times K_v(\sqrt{2gH})$$

$$d = \sqrt{\frac{4Q}{\pi n_j} \times \frac{1}{C_v\sqrt{2gH}}}$$

$$d = \frac{0.54\sqrt{\frac{Q}{n_j}}}{H^{\frac{1}{4}}} \tag{5.53}$$

5.5.1.3 Casing

It does not perform any hydraulic function which is just essential to stop splashing of water, direct it to the tail race, and also serves as an accident prevention component. The typical dimensions of casing for Pelton turbine with respect to jet diameter (d) as shown in Fig. 5.15.

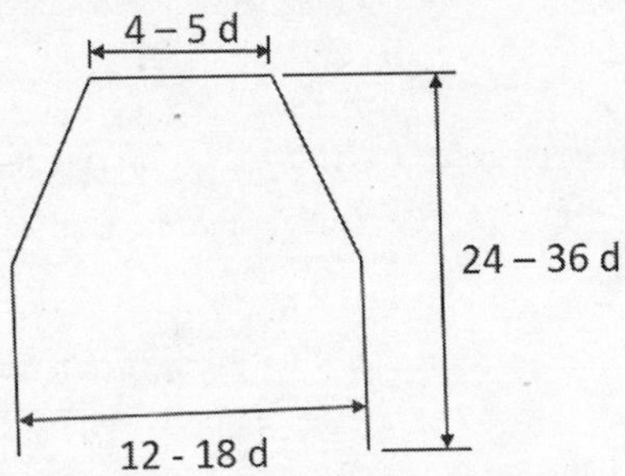

Fig. 5.15: Casing of Pelton turbine

5.5.1.4 Control valves

They are provided to stop the flow during emergency conditions.

5.5.1.5 Shaft with bearing

The shaft is provided to transmit the power from the turbine runner to the end use (Generator) while, bearings are provided to support the shaft.

EXAMPLE 5.5

Example 5.5 Pelton turbine installed at a site produces 1000 kW. It works under the head of 350. Assuming suitable values of coefficient of velocity, speed ratio and overall efficiency find the least diameter of runner, number of jets and flow rate.

Solution

Given data:

The capacity of the Pelton turbine

Power, $P = 1000$ kW

Net head, $H = 350$ m

Assuming values

Coefficient of velocity, $K_v = 0.98$

Speed ratio, $K_u = 0.46$

Overall efficiency, $\eta_o = 0.98$

$$u = 0.46 \times V_{jet}$$

$$u = 0.46 \times 0.98 \times \sqrt{2gH}$$

$$\frac{\Pi DN}{60} = 0.46 \times 0.98 \times \sqrt{2 \times 9.81 \times 350}$$

$$DN = 713.45$$

Assume speed, $N = 1500\ rpm$

Diameter of runner, $D = \frac{713.45}{N}$

$$= \frac{713.45}{1500}$$

$D = 0.475$ m. **Ans.**

Efficiency $\eta = \frac{P}{\rho gQH}$

Power $P = \eta_o \rho gQH$

$$1000 \times 10^3 = 0.9 \times 1000 \times 9.81 \times Q \times 350$$

Flow rate, $Q = 0.326$ m^3/sec. **Ans.**

Number of jets, n

$$Q = A \times V_{jet} \times n$$

$$Q = \frac{\Pi}{4} \times d_{jet}^2 \times 0.98 \times \sqrt{2gH} \times n$$

$$0.326 = \frac{\Pi}{4} \times d_{jet}^2 \times 0.98 \times \sqrt{2 \times 9.81 \times 350} \times n$$

To find d_{jet} assume, $m = \frac{D}{d} = 10$

$$d = 0.0475\ m$$

$$0.326 = \frac{\pi}{4} \times 0.0475^2 \times 0.98 \times \sqrt{2 \times 9.81 \times 350} \times n$$

$$n = 2.4$$

Conclusion: The number of jets should be in the range of 2 or 3. **Ans.**

Example 5.6 Calculate the power given by the jet of water to the runner of a Pelton wheel which is having tangential velocity as 21 m/s. The net head on the turbine is 52 m and discharge through the jet water is 0.05 m^3/s. The side clearance angle is 15° and take C_v = 0.98.

Solution

Given data:

Tangential velocity of wheel, $u = u_1 = u_2 = 21$ m/s

Net head, $H = 52$ m

Discharge, $Q = 0.05\ m^3/s$

Side clearance angle, $\beta_1 = 15°$

Co – efficient of velocity, $C_v = 0.98$

Velocity of the jet, $v_1 = C_v \times \sqrt{2gH}$

$$= 0.98 \times \sqrt{2 \times 9.81 \times 52}$$

$$= 31.302\ m/s$$

From inlet triangle, $v_{u_1} = v_1 = 31.302\ m/s$

$$w_1 = v_{u_1} - u_1$$

$$= 10.302\ m/s$$

From outlet velocity triangle, we have

$$w_1 = w_2 = 10.302\ m/s$$

$$w_2 \cos\beta_1 = 10.302 \cos 15° = 9.95\ m/s$$

$$v_{u_2} = u_2 - w_2 \cos\beta_1 = 21 - 9.95 = 11.05\ m/s$$

Also as β_2 is an obtuse angle, the work done per second on the runner,

$$= \rho a v_1 [v_{u_1} - v_{u_2}] \times u = \rho QS[v_{u_1} - v_{u_2}] \times u$$

$$= 1000 \times .05 \times [31.302 - 11.05] = 1012.6\ Nm/s$$

∴ Power given to the runner in kW = $\frac{Work\ done\ per\ second}{1000}$

$= \frac{1012.6}{1000} = 1.012$ kW. **Ans.**

EXAMPLE 5.6

Example 5.7 A Pelton wheel is to be designed to develop 1,000 kW at 400 r.p.m. It is to be supplied with water from a reservoir whose level is 250 m above the wheel through a pipe 900 m long. The pipeline losses are to be 5% of gross head. The co-efficient of friction is 0.005. The bucket speed is to be 0.46 of jet speed and efficiency of wheel is 85%. Calculate the pipeline diameter, jet diameter, and wheel diameter.

EXAMPLE 5.7

EXAMPLE 5.7

Solution

Given data:

The capacity of the Pelton turbine

Power, $P = 1000\ kW$

Speed $N = 400\ rpm$

Gross head, $H_g = 350$

Coefficient of friction $f = 0.005$

$$u = 0.46 \times V_{jet}$$
$$L = 900\ m$$
$$h_L = 5\% \times H_g$$
$$h_L = 5\% \times 350$$
$$h_L = 12.5\ m$$

Net head, $H = H_g - h_L$

$$H = 250 - 12.5$$
$$H = 237.5\ m$$
$$u = 0.46 \times V_{jet}$$
$$\frac{\Pi DN}{60} = 0.46 \times \sqrt{2gH}$$
$$\frac{\Pi \times D \times 400}{60} = 0.46 \times \sqrt{2 \times 9.81 \times 237.5}$$
$$D = 1.469\ m$$

Efficiency $\eta = \frac{P}{\rho g Q H}$

$$0.85 = \frac{1000 \times 10^3}{1000 \times 9.81 \times Q \times 237.5}$$

Total flow rate $Q = 0.5\ m^3/_s$

$$Q = A \times V_{jet} \times n_{jet}$$
$$Q = \frac{\Pi}{4} \times d_{jet}^2 \times 0.98 \times \sqrt{2gH} \times n_{jet}$$

Jet ratio $m = \frac{D}{d_{jet}} = 10$

$$d_{jet} = \frac{1.469}{10}$$

$d_{jet} = 0.1469$ m. **Ans.**

Head loss due to friction, $h_L = \frac{4fLV^2}{2gD_p}$

$$Q = \frac{\Pi}{4} \times D_p^2 \times V$$
$$V = \frac{4Q}{\Pi D_p^2}$$
$$h_L = \frac{4fL \times 4^2 \times Q^2}{2g\Pi^2 D_p^5}$$
$$12.5 = \frac{4 \times 0.005 \times 900 \times 4^2 \times 0.5^2}{2 \times 9.81 \times \Pi^2 \times D_p^5}$$
$$D_p^5 = 0.0297$$

$D_p = 0.495$ m. **Ans.**

5.6 FRANCIS TURBINE

Francis turbine is basically a reaction turbine where the runner receives the water under pressure in a radial inwards direction and discharges it in axial direction. Water imparts its energy gradually while flowing from inlet to the outlet of the runner due to reaction created between the water and runner. The Francis turbine like other reaction turbines is a closed turbine which is also known as *variable pressure turbine.* The water pressure is continuously varying from maximum at the entrance to the minimum at the exit of the runner. The basic components of Francis turbine are as shown in Fig. 5.16.

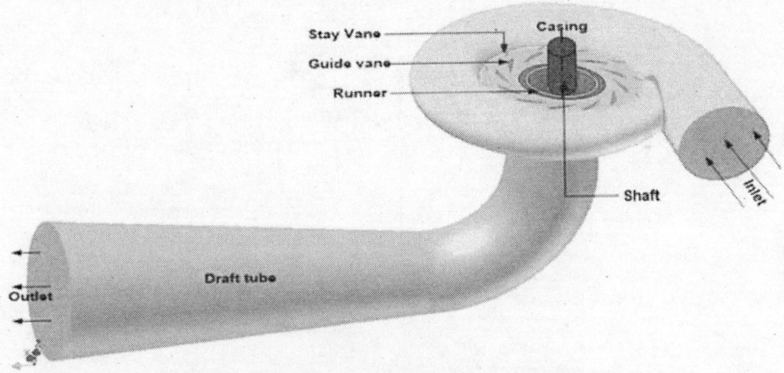

Fig. 5.16: Basic components of Francis turbine

5.6.1 Basic Components of Francis Turbine

Followings are the basic components of typical Francis turbine:

(i) Spiral casing

(ii) Guide vanes

(iii) Runner

(iv) Draft tube

5.6.1.1 Spiral casing

As Francis turbine is a closed turbine which requires proper casing it is therefore, a spiral casing (volute casing) is provided. The schematic diagram of spiral casing is shown in Fig. 5.17. The function of spiral casing is to distribute the water with constant velocity around the runner through guide vanes provided in the turbine.

For given head (H) and discharge (Q), the basic design of volute casing is described as follows:

The quantity of water flowing per second at the entrance of the casing can be expressed as

$$Q = \frac{\pi}{4} d_i^2 v \tag{5.54}$$

Where, v is the spouting velocity which expressed as

$$v = \sqrt{2gH} \tag{5.55}$$

and d_i is the inlet diameter of spiral/volute casing which is kept approximately equal to the diameter of penstock (inlet pipe) and is determined by the following Eq. 5.56.

$$d_i = \sqrt{\frac{4Q}{\pi v_i}} \tag{5.56}$$

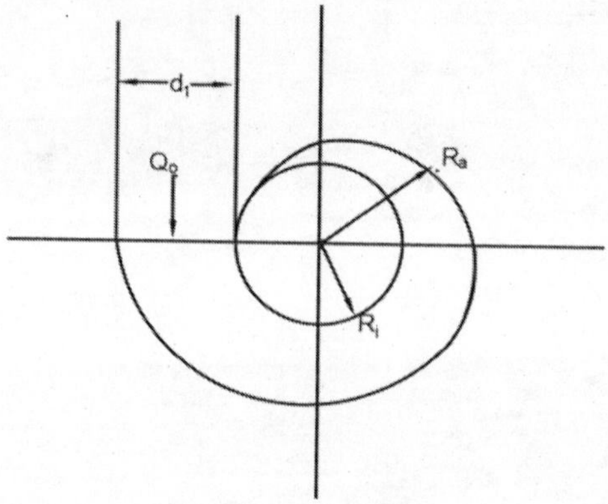

Fig. 5.17: Spiral Casing

Other dimensions as shown in Fig. 5.17 are determined as follows

$$R_a = R_i + \frac{\theta}{2\pi} d_i \tag{5.57}$$

$$Q_\theta = Q\,\frac{\theta}{2\pi} \tag{5.58}$$

Where, R_i is the inner circle radius,

R_a is the volute radius,

Q_θ is the discharge at different section of the casing.

A full spiral is generally recommended for high head and semi-spiral for low head installation.

5.6.1.2 Guide vanes

Guide vanes in Francis turbines are used to guide the water from casing to the runner. A guide vane has hydrofoil profile as shown in Fig. 5.18. Number of guide vanes are attached with the rim around the runner and can be adjusted by the governor as per the requirement of the water to the turbine. The other function of guide vanes is to convert the potential energy into kinetic energy partially. The flow area between the two guide vanes at different positions is shown in Fig. 5.18.

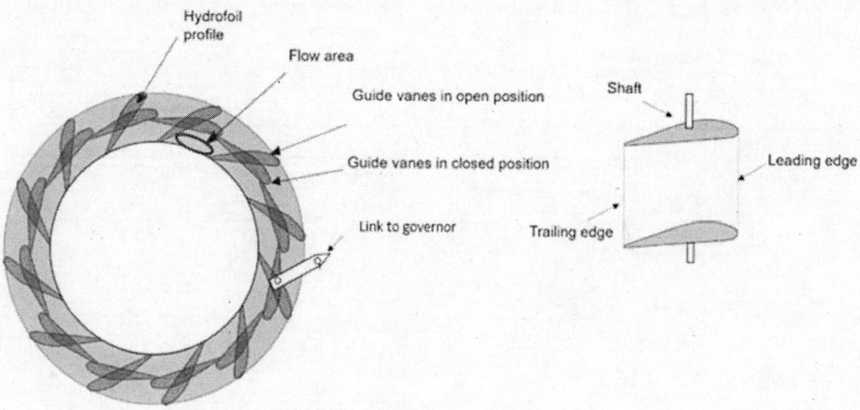

Fig. 5.18: Guide vanes

The basic design of guide vanes is described as follows:

Assuming U_o is the hypothetical peripheral velocity corresponding to the diameter (D_o) measured by the distance between the trailing edge of two opposite guide vanes which can be expressed by the Eq. 5.59.

$$U_o = \frac{\pi D_o N}{60} \tag{5.59}$$

$$U_o = K_{Uo}\sqrt{2gH} \tag{5.60}$$

Where, N is the rotational speed

K_{U_o} is the speed ratio (in the range of 0.7 to 1.34) considered for design of the guide vanes,

H is the head.

Further, D_o can be determined using the following expression as given in Eq. 5.61.

$$D_o = \frac{60 K_{Uo}\sqrt{2gH}}{\pi N} \tag{5.61}$$

Accordingly, the length (L) of guide vane is determined by the Eq. 5.62

$$L = 0.3\ D_o \tag{5.62}$$

The width of guide vane B_0 will be equal to the breadth of Francis runner.

Further, the number of guide vanes (Z) depends on the runner diameter and specific speed of the turbine which varies from 8 to 24.

In practise the number of guide vanes can be selected from Table 5.1.

5.6.1.3 Francis runner

Francis turbine is mixed flow reaction turbine where water enters over the runner radially and leaves the water in axial direction i.e., parallel to the shaft. It consists of number of fixed runner vanes between two discs. One disk is fitted to the shaft while another is opened to allow the water flow to the draft tube as shown in Fig. 5.19.

Table 5.1: Selection of number of guide vanes

N_s	$Z = 8$	10	12	14	16	18	20	24
Upto 200	Value of D_o in mm							
	Upto 250	250 400	400 600	600 800	800 1000	1000 1250	1250 1700	More 1700
More than 200	Upto 300	300 450	450 750	750 1050	1050 1350	1350 1700	1700 2100	More 2100

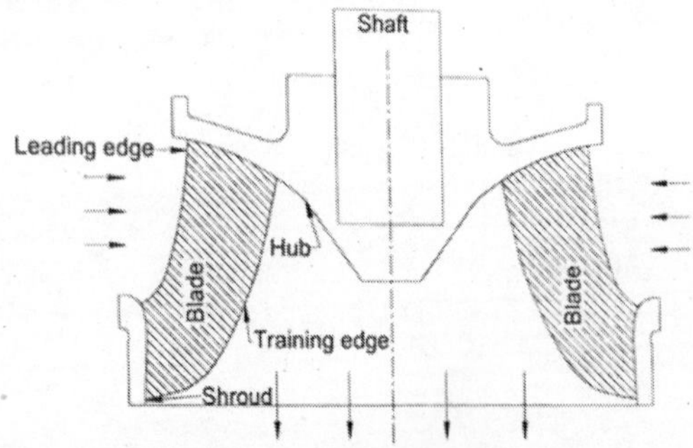

Fig. 5.19: Schematic diagram of Francis runner

Since Francis turbines cover a wider range of head and discharge in practise it is therefore, Francis runners are classified as slow, medium and fast depending upon specific speed.

Francis runners are classified as;

(i) Fast runner suitable for high head and low discharge having larger diameter and low breadth of the runner,

(ii) Medium runner suitable for medium head and medium discharge having diameter and breadth accordingly, and

(iii) Slow runner suitable for low head and high discharge having small diameter and larger breadth.

The velocity triangles for different runners are shown in Fig. 5.20.

In order to design the Francis runner following working proportions are considered;

(i) The ratio of the width B of the runner to the diameter D_o of the runner ranges from 0.10 to 0.45.

(ii) The ratio of the velocity of flow v_f to the spouting velocity which is known as flow ratio (Eq. 5.63) and its range varies from 0.15 to 0.30.

$$\psi = \frac{v_f}{\sqrt{2gH}} \tag{5.63}$$

(iii) The speed ratio K_u is defined as the ratio of peripheral velocity of runner to the spouting velocity (Eq. 5.64) and its range varies from 0.60 to 0.90.

$$K_u = \frac{u}{\sqrt{2gH}} \tag{5.64}$$

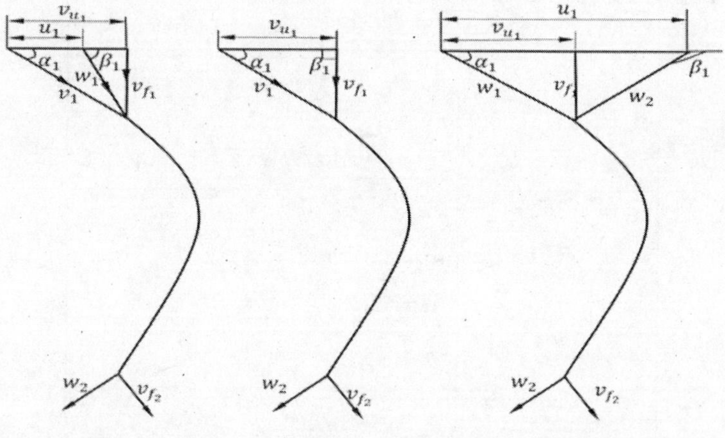

Fig. 5.20: Velocity triangle diagram for each runner (a) Slow (b) Medium (c) Fast

Using the above working proportions the runner sizes can be determined by the following steps are given below;

Step 1: Determine the specific speed for the given head (H) and capacity of the turbine (P) and assuming speed (N) to select the type of runner and accordingly the values of different parameters $(\frac{B}{D_o}, \psi, and\ K_u)$ from the Table 5.2.

Step 2: Determine the runner diameters and number of runner vanes by considering the given head (H) and its speed (N) by selecting the $K_u$value using the expression as given in Eq. 5. 65.

$$K_u \sqrt{2gH} = \frac{\pi D_o N}{60} \tag{5.65}$$

$$D_o = \frac{60}{\pi N} K_u \sqrt{2gH} \tag{5.66}$$

Runner inner diameter $D_i = D_o/2$

Number of runner vanes $Z_1 = Z + 1$

Step 3: Determine the runner width by considering the given flow rate which is equated as the product of cross-sectional area (width × circumferential area) and flow velocity and expressed as;

$$Q = \pi D_o B v_f \tag{5.67}$$

If the blockage effect due to blade thickness is taken, then

$$Q = (\pi\, D_o - Z_1 t) B\, v_f \tag{5.68}$$

Where, (Z_1) are the number of runner vanes, (B) is width of runner and (t) is the thickness of vane.

Table 5.2: Values of different parameters for various Francis runners.

Type of runner	Values
Slow runner	$N_s = 60$ to 120 $\alpha_1 = 15^0$ to 25^0 $K_{u_1} = 0.62$ to 0.68 $\beta_1 = 90^0$ to 120^0 $\frac{B}{D_o} = 0.03$ to 0.04
Medium runner	$N_s = 120\ to\ 180$ $\alpha_1 = 25^0$ to $32\,\frac{1}{2}^0$ $K_{u_1} = 0.68$ to 0.72 $\beta_1 = 90^0$ $\frac{B}{D_o} = 0.125\ to\ 0.25$ $\beta_1 = 60^0\ to\ 90^0$
Fast runner	$N_s = 180$ to 400 $\alpha_1 = 32\,\frac{1}{2}^0\ to\ 37\,\frac{1}{2}^0$ $K_{u_1} = 0.72\ to\ 0.9$ $\beta_1 = 60 + 90^0$ $\frac{B}{D_o} = 0.25\ to\ 0.45$ $\alpha_2 = 90^0$ for outlet velocity triangle

5.6.1.4 Draft Tube

In reaction turbine which are basically closed turbine and gross head is considered across the turbine as difference between upstream and downstream water levels. The draft tube facilitates to install the turbine at any level without losing the head. Further, it also acts as an energy recuperater to recover the kinetic energy going as waste at the exit of the runner.

In order to discuss the working principle of draft tube Fig. 5.21 can be refereed where required parameters are shown.

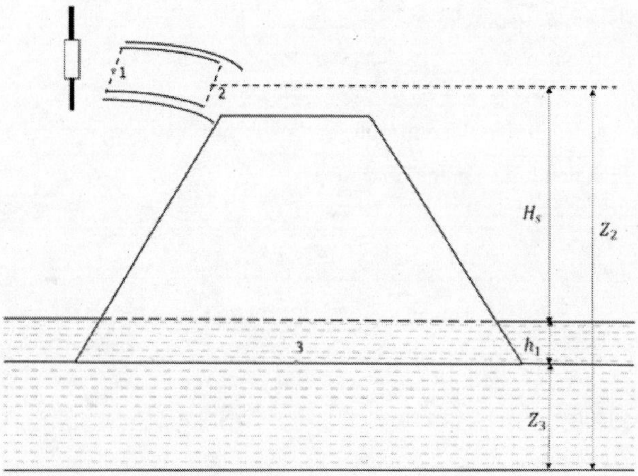

Fig. 5.21: Draft tube of Francis turbine

The different locations are indicated as runner entrance (1), runner centre (2) and draft tube exit (3)

Applying Bernoulli's equation between locations 2 and 3 and we get

$$\frac{p_2}{\gamma}+\frac{v_2^2}{2g}+z_2=\frac{p_3}{\gamma}+\frac{v_3^2}{2g}+z_3+h_f \tag{5.69}$$

Where P_2, P_3 and V_2, V_3 are the pressures and velocities at point 2 and 3 respectively and h_f is the loss of head in the draft tube

Now,

$$\frac{p_2}{\gamma}=\frac{p_3}{\gamma}-(z_2-z_3)-\left(\frac{v_2^2-v_3^2}{2g}\right)+h_f \tag{5.70}$$

Where, $\frac{p_3}{\gamma}=\frac{p_a}{\gamma}+h_1$

Therefore,

$$\frac{p_2}{\gamma}=\frac{p_a}{\gamma}+h_1-(z_2-z_3)-\left(\frac{v_2^2-v_3^2}{2g}\right)+h_f \tag{5.71}$$

Since $(Z_2-Z_3-h_1)=H_s$, the height of runner exit above the tailrace level, thus

$$\frac{p_2}{\gamma}=\frac{p_a}{\gamma}-\left[H_s+\frac{v_2^2-v_3^2}{2g}\right]+h_f \tag{5.72}$$

Where,

the pressure at the runner exit is suction pressure that is below atmospheric pressure,

H_s is known as static suction head,

$\left[\frac{v_2^2 - v_3^2}{2g}\right]$ is known as dynamic suction head,

Geometrically h_f is expressed as $h_f = k\left[\frac{v_2^2 - v_3^2}{2g}\right]$,

h_f is the head loss due to friction in the draft tube and

k is the coefficient of friction losses in the draft tube

Then

$$\frac{p_2}{\gamma} = \frac{p_a}{\gamma} - \left[H_s + (1-k)\frac{v_2^2 - v_3^2}{2g}\right] \tag{5.73}$$

The efficiency of a draft tube is defined as;

$$\eta_d = \frac{\text{Actual regain of pressure head}}{\text{velocity head at entrance of draft tube}}$$

The actual regains of pressure head

$$\left[\frac{v_2^2 - v_3^2}{2g}\right] - h_f = (1-k)\left[\frac{v_2^2 - v_3^2}{2g}\right]$$

$$\eta_d = \frac{\left[\frac{v_2^2 - v_3^2}{2g}\right] - h_f}{\frac{v_2^2}{2g}} \tag{5.74}$$

Types of draft tubes

There are three different types of draft tubes are available as shown in Fig. 5.22.

(i) Straight divergent Tube

(ii) Moody's spreading Tube

(iii) Elbow type draft tube

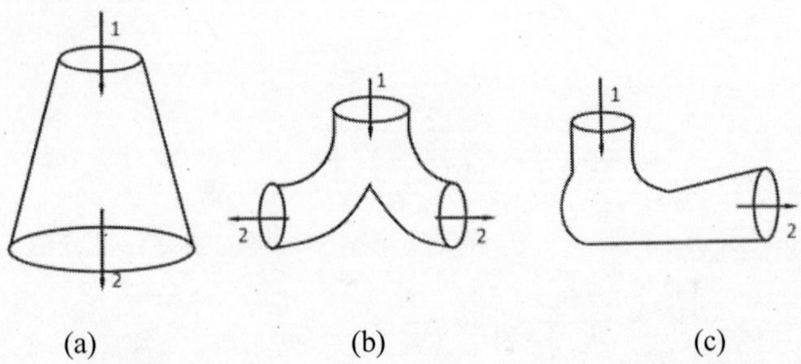

Fig. 5.22: Types of draft tube (a) Straight divergent (b) Moody's spreading tube (c) Elbow type with a circular inlet and a rectangular outlet section

Among all these three draft tubes, straight divergent and Moody's draft tube are considered as most efficient.

Example 5.8 Design a Francis turbine runner with the following data. Net head H = 70 m, speed N = 760 rpm, output power P = 320 kW, Hydraulic efficiency (η_h) = 95%, overall efficiency (η_o) = 87%, flow ratio (ψ) = 0.16, breath ratio = 0.1, inner diameter is (1/2) outer diameter. Also assume 5% of circumferential area of the runner to be occupied by the thickness of the vanes. Velocity of flow remains constant throughout and flow is radial at exit.

Solution

Given data:

$H = 70$ m,

$N = 760$ r.p.m

$P = 320$ kW

$\eta_h = 95\%$

$\eta_o = 87\%$

$\psi = 0.16$

$\frac{B}{D_0} = 0.1$

$\frac{D_i}{D_0} = 0.5$

Efficiency is given by

$$\eta_o = \frac{P}{\rho g Q H}$$

$$0.87 = \frac{320 \times 10^3}{9810 \times Q \times 70}$$

Discharge, $Q = 0.535$ m^3/sec

Flow velocity is given by

$$v_f = \psi\sqrt{2gH}$$

$$= 0.16\sqrt{2 \times 9.81 \times 70} = 5.9 \text{ m/sec}$$

Discharge also can be expressed as

$$Q = (k\pi n D^2)\, v_f$$

$$0.535 = 0.95 \times \pi \times 0.1 \times D^2 \times 5.9$$

Outer diameter of runner, $D_o = 551.35$ mm

Breadth ratio $= \frac{B}{D_0} = \frac{B}{551.35}$

$B = 55.135$ mm

Tangential velocity, $u_1 = \frac{\pi D_0 N}{60}$

$$= \frac{\pi \times 0.55 \times 760}{60} = 21.93 \text{ m/sec}$$

EXAMPLE 5.8

EXAMPLE 5.8

Hydraulic efficiency, $\eta_h = \frac{v_{u_1} u_1}{gH}$

$$0.9 = \frac{v_{u_1} \times 21.93}{9.81 \times 70}$$

$$v_{u_1} = 29.75 \text{ m/sec}$$

From inlet velocity triangle,

$$\tan \alpha_1 = \frac{v_{f_1}}{v_{u_1}} = \frac{5.9}{29.75} = 0.1983$$

$$\alpha_1 = 11.216°$$

$$\tan \beta_1 = \frac{v_{f_1}}{v_{u_1} - u_1} = \frac{5.9}{29.75 - 21.93} = 0.754$$

$$\beta_1 = 37.03°$$

Now $D_i = \frac{D_o}{2} = 275.68$ mm

$$\frac{u_1}{D_i} = \frac{u_2}{D_o};\ u_2 = 10.963 \text{ m/sec}$$

From outlet velocity triangle,

$$\tan \alpha_2 = \frac{v_f}{u_2} = \frac{5.9}{10.963} = 0.538$$

$$\alpha_2 = 28.28°$$

EXAMPLE 5.9

Example 5.9 A Francis turbine with an overall efficiency of 77% is required to produce 150 kW power. It is working under a head of 7 m. The peripheral velocity $0.3\sqrt{2gH}$ and the radial velocity of flow at inlet is $0.9\sqrt{2gH}$. The wheel run at 160 rpm and the hydraulic losses in the turbine are 20% of the available energy. Assuming radial discharge, determine:

(a) The guide blade angle
(b) The wheel vane angle at inlet
(c) Diameter of the wheel at inlet

Solution

Given data:

Overall efficiency, $\eta_o = 77\% = 0.77$

Power Produced, $S.P = 150$ kW

Head $H = 7$ m

Peripheral velocity, $u = 0.3\sqrt{2gH} = 3.516$ m/sec

Velocity of flow at inlet, $v_f = 0.9\sqrt{2gH} = 10.55$ m/sec

Speed $N = 160$ r.p.m.

Hydraulic losses = 20% of available energy

Hydraulic efficiency as,

$$\eta_h = \frac{H - 0.2\,H}{H} = \frac{0.8\,H}{H}$$

$$0.8 = \frac{V_u\, u}{gH}$$

Whirl velocity, $v_u = 15.62$ m/s

(a) The guide blade angle (α), from inlet triangle

$$\tan \alpha_1 = \frac{v_f}{v_u} = \frac{10.55}{15.62} = 0.675$$

$\alpha_1 = 34.02$. **Ans.**

(b) The wheel vane angle at inlet

$$\tan \alpha_2 = \frac{v_f}{v_u - u} = \frac{10.55}{15.62 - 3.516} = 0.8716$$

$\alpha_2 = 41.075$. **Ans.**

(c) Diameter at inlet (D)

$$u = \frac{\pi D N}{60}$$

$$D = \frac{60 \times u}{\pi \times N}$$

$$= \frac{60 \times 3.516}{\pi \times 160} = 0.419 \text{ m. } \textbf{Ans.}$$

EXAMPLE 5.9

5.7 KAPLAN TURBINE

Kaplan turbines are also reaction type and axial flow turbines where water flow over the runner purely in the axial direction i.e., parallel to the shaft from inlet to outlet. Under low head and high discharge these turbines are suitable. Basic components of axial flow turbines are similar to Francis turbine, where the guide vanes and runner vanes are generally called as wicket gates and runner blades, respectively (Fig. 5.23).

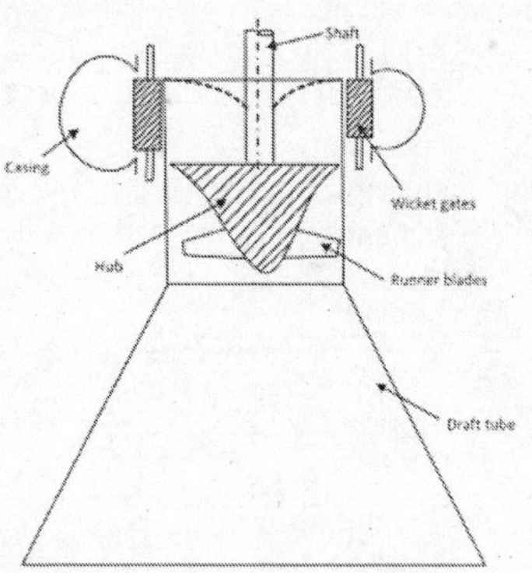

Fig. 5.23: Kaplan turbine

The axial flow turbine has adjustable wicket gates and it may or may not have adjustable blades accordingly turbine known as Kaplan or propeller turbine respectively. The propeller turbine having fixed runner blades is recommended if discharge variation is not there at the sites.

However, Kaplan turbine having adjustable blades should be recommended for a site where discharge variation at site is high. The orientation of the shaft may be vertical or horizontal depending upon the site conditions. Kaplan turbines are further classified in different categories depending upon their shape which are discussed in the following sub section.

5.7.1 Types of Kaplan Turbine

5.7.1.1 Tubular turbine

As Kaplan turbines are axial flow where water is flowing parallel to the shaft which derives the generator it becomes necessary to protect the generator from water. Accordingly, the runner is encased in tubes hence the turbine named as tubular turbine and the generator is placed outside the tube. The most popular turbine is S-type tubular turbine in this configuration where the tube shape is 'S' as shown in Fig. 24.

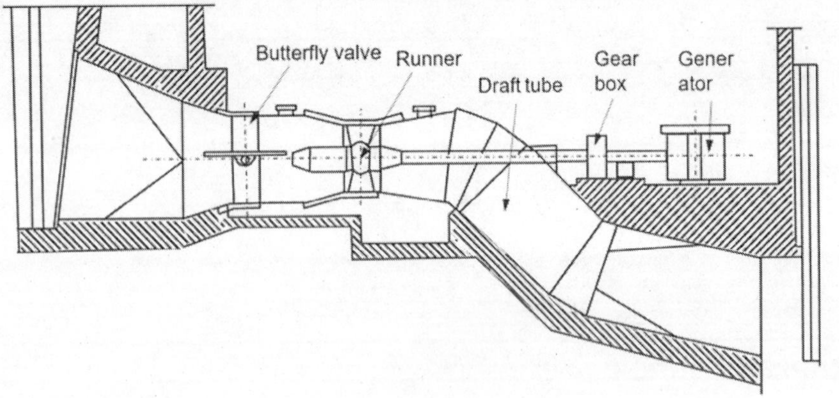

Fig. 5.24: S-type tube turbine

5.7.1.2 Bulb Turbine

Nowadays bulb turbine is considered the most efficient turbine for low head site. In bulb turbine generator connected to the runner through gearbox is encased in a bulb. The bulb unit is placed horizontally completely submerged in water passage as shown in Fig. 5.25.

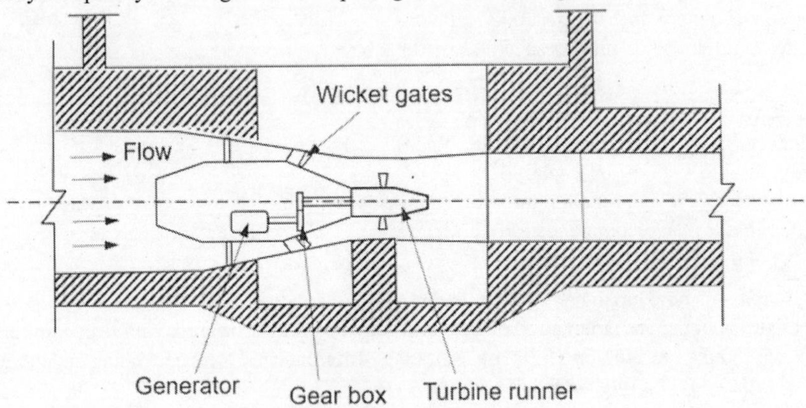

Fig. 5.25: Bulb turbine

5.7.1.3 Rim or Straflo Turbine

A Rim turbine looks like a bulb turbine but a separate alternator is not required. The alternator rotor is mounted on the periphery of the turbine runner blades and stator is put inner surface of the tube as shown in Fig. 5.26. However, research is still going on for the commercialization of this turbine.

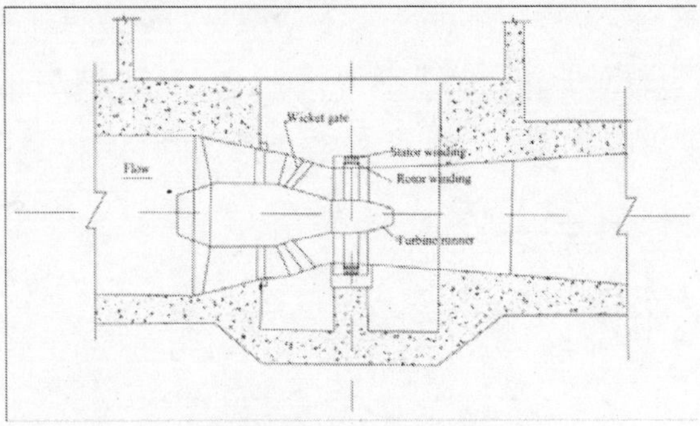

Fig. 5.26: Straflo turbine

5.7.2 Basic Components of Kaplan Turbine

Following are the basic components of typical Kaplan turbine:

(i) Runner with blades

(ii) Wicket gates

(iii) Draft tube

(iv) Casing

5.7.2.1 Runner with blades

Runner of a Kaplan turbine consists of number of blades with a hub which is keyed to the turbine shaft. The blades may be adjustable according to the water flow variation over the runner to maintain the part load efficiency.

Depending on head (H) and discharge (Q), the design of Kaplan runner follows the same steps of Francis runner design as discussed under the section 5.6.1.3. Following working proportions of Kaplan runner are considered as given below:

(i) The ratio of the hub diameter D_h to the runner diameter D_r of the runner ranges from 0.3 to 0.65.

(ii) The flow ratio, ψ range varies from 0.35 to 0.60.

$$\psi = \frac{v_f}{\sqrt{2gH}} \tag{5.75}$$

(iii) The speed ratio, K_u range varies from 0.90 to 3.0.

$$K_u = \frac{u}{\sqrt{2gH}} \tag{5.76}$$

Using the above working proportions the runner sizes can be determined by the following steps given below;

Step 1: Determine the specific speed for the given head (H) and capacity of the turbine (P) by assuming speed (N) of the turbine to select the values of different parameters ($\frac{D_h}{D_r}, \psi, and\ K_u$).

Step 2:

Diameter of the runner is determined by following steps

$$\text{Discharge,} \quad Q = \frac{\pi}{4}(D_r^2 - D_h^2)v_f \tag{5.77}$$

$$\frac{4Q}{\pi v_f} = D_r^2\left[1 - \left(\frac{D_h}{D_r}\right)^2\right]$$

$$D_r^2 = \frac{4Q}{\pi v_f[1 - k^2]}$$

Where, $k = \frac{D_h}{D_r}$

$$D_r^2 = \frac{4Q}{\pi\psi\sqrt{2gH}[1 - k^2]}$$

$$D_r = \frac{0.536}{H^{1/4}}\sqrt{\frac{Q}{(1 - k^2)\psi}} \tag{5.78}$$

The Values of $\frac{D_h}{D_r}, \psi, and\ K_u$ can be selected from the Table 5.3 as per the specific speed.

Table: 5.3: Values of different parameters for design of Kaplan runner

Specific Speed N_s	**400**	**600**	**800**	**1000**
Hub to runner diameter ratio, $\frac{D_h}{D_r}$	0.3 - 0.35	0.35 - 0.45	0.45 - 0.55	0.55 - 0.65
Flow ratio, ψ	0.35 - 0.38	0.38 - 0.42	0.42 - 0.52	0.52 - 0.60
Speed ratio, K_u	0.90 - 0.95	0.95 - 1.0	1.0 - 2.0	2.0 - 3.0

In case of Kaplan turbine, runner of blades varies from 3 to 8 depending on the specific speed.

Further, the work done by water and efficiency of the Kaplan runner is explained with the velocity triangle is shown in Fig. 5.27.

Work done by the water per sec

$$W = m(v_{u1} + v_{u2})u$$
$$= \rho Q(v_{u1} + v_{u2})u \qquad (5.79)$$

For axial for turbine ($u = u_1 = u_2$)

Efficiency is given by

$$\eta = \frac{P_{output}}{P_{input}}$$
$$\eta = \frac{\rho Q(v_{u1} + v_{u2})u}{\rho g Q H}$$
$$\eta = \frac{(v_{u1} + v_{u2})u}{gH} \qquad (5.80)$$

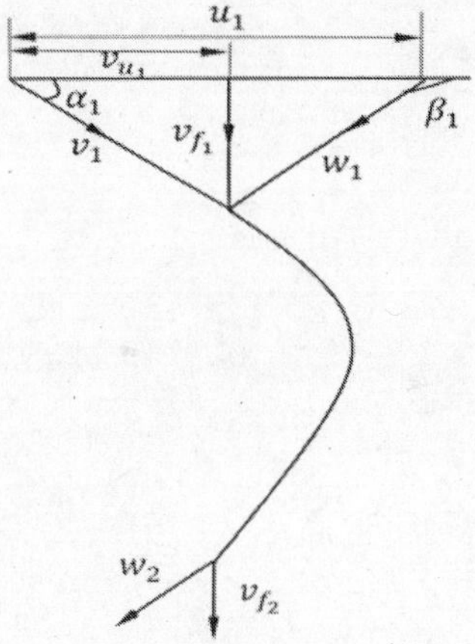

Fig. 5.27: Velocity Triangle for Kaplan Turbine

5.7.2.2 Wicket gates

Wicket gates are provided to regulate the flow as per the load variations so to run the turbine at constant speed. The wicket gate diameter (D_o) can be calculated by using following Eq. 5.81

$$D_o = D_r \times K_{u1} \qquad (5.81)$$

Where, K_{u1} is the speed ratio considered from the given Table 5.4.

Table: 5.4 Selection of speed ratio for the design of wicket gate

N_s	400	600	800	1000
K_{u1}	1.12	1.06	1.01	1

Accordingly, the width of wicket gate (B) is determined by the Eq. 5.82.

$$B = 0.4\, D_r \qquad (5.82)$$

Further, the number of wicket gates (Z) depends on the wicket diameter which varies from 8 to 24. In practise the number of wicket gates can be selected from Table 5.5.

Table 5.5: Selection of number of guide vanes

D_o (in mm)	Upto 300	300-450	450-750	750-1200	1200-1600	1600-2200	2200-4000	7400
Z	8	10	12	14	16	18	20	24

5.7.2.3 Draft tube

The design and theory of draft tube is similar as discussed under Francis turbine section.

Example 5.10 A Kaplan turbine produces 50000 kW under a head of 20 m with an overall efficiency of 88%. Consider the value of speed ratio as 1.5, flow ratio as 0.45 and the hub diameter as 0.4 times the outer diameter, calculate the diameter and speed of turbine.

Solution

Given data:

Power, P = 50000 kW

Head, H = 20 m

Overall efficiency, η_o = 88%

Speed ratio, K_u = 1.5

Flow ratio, ψ = 0.45

Hub diameter, $D_h = 0.4D_r$

Efficiency is given by

$$\eta_o = \frac{P}{\rho g Q H}$$

$$0.88 = \frac{50\,000 \times 10^3}{1000 \times 9.81 \times Q \times 20}$$

$$Q = 289 \text{ m}^3/\text{sec}$$

$$Q = \frac{\pi}{4}\left(D_r^{\,2} - D_h^{\,2}\right)\psi\sqrt{2gH}$$

$$289 = \frac{\pi}{4} \times D_r^{\,2}(1 - 0.4^2) \times 0.4 \times \sqrt{2 \times 9.81 \times 20}$$

Diameter of turbine

D_r = 7.43 m **Ans.**

Peripheral velocity is given by

$$u = K_u\sqrt{2gH}$$

$$u = 1.5 \times \sqrt{2 \times 9.81 \times 20}$$

$$= 29.71 \text{ m/sec.}$$

$$u = \frac{\pi D N}{60}$$

$$29.71 = \frac{\pi \times 7.43 \times N}{60}$$

Speed,

N = 73.63 r.p.m. **Ans.**

EXAMPLE 5.10

Example 5.11 The hub diameter of a Kaplan turbine, working under a head of 10 m, is 0.40 times of the diameter of the runner. The turbine is running at 120 r.p.m. If the vane angle of the extreme edge of the runner at outlet is 15^0 and flow ratio is 0.55, calculate

(i) Runner diameter

(ii) Hub diameter

(iii) Discharge

The velocity of whirl at outlet is given as zero.

(i) Runner diameter

$$\psi = \frac{v_{f_1}}{\sqrt{2gH}}$$

$$0.55 = \frac{v_{f_1}}{\sqrt{2 \times 9.81 \times 10}}$$

$$v_{f_1} = 7.703 \text{ m/sec}$$

From the outlet velocity triangle, $v_{u_2} = 0$

$$\tan \beta_2 = \frac{v_{f_2}}{u_2} = \frac{v_{f_1}}{u_1}$$

$$\tan 15^0 = \frac{7.703}{u_1}$$

$$u_1 = u_2 = 28.74 \text{ m/sec}$$

$$u_1 = \frac{\pi D_r N}{60}$$

$$28.74 = \frac{\pi \times D_r \times 120}{60}$$

$$D_r = 4.574 \text{ m. } \textbf{Ans.}$$

(ii) Hub diameter

$$D_h = 0.4\, D_r$$

$$D_h = 1.829 \text{ m. } \textbf{Ans.}$$

(iii) Discharge

$$Q = \frac{\pi}{4}\left(D_r^{\,2} - D_h^{\,2}\right)\psi\sqrt{2gH}$$

$$Q = \frac{\pi}{4} \times 4.574^2(1 - 0.4^2) \times 0.55 \times \sqrt{2 \times 9.81 \times 10}$$

$$Q = 106.33 \text{ m}^3\text{/sec}$$

EXAMPLE 5.11

5.8 SIMILARITY LAW

Similarity law says that two machines can be called similar if they are geometrically, kinematically, and dynamically similar. In practise the prototype of water turbines is not possible to test in the laboratory. The efficiency of prototype is predicted based on the efficiency of reduced size machine (model) similar to the prototype which is possible to test in the laboratory. The similarity law is discussed in the following section;

5.8.1 Geometrical Similarity

Two machines are said to be geometrically similar if the ratio of physical dimensions of two components is similar. In case of a Francis turbine, ratio of runner width to the runner outer diameter is considered to show the geometrical similarity as follows

$$\left(\frac{B}{D_1}\right)_m = \left(\frac{B}{D_1}\right)_p \tag{5.86}$$

where, m stands for model and p stands for prototype.

B is runner width and

D_1 is outer runner diameter

Similarly, the geometrical similarity of Pelton turbine will be expressed by the ratio of runner diameter to jet diameter and for Kaplan turbine by the ratio of hub diameter to runner diameter.

5.8.2 Kinematic Similarity

As per the kinematic similarity the ratio of two velocity components of one machine should be equal to that of another machine. In case of Francis turbine, the ratio of flow velocity to peripheral velocity is considered to show the kinematic similarity as follows

$$\left(\frac{v_f}{u}\right)_m = \left(\frac{v_f}{u}\right)_p \tag{5.87}$$

Similarly, the kinematic similarity of Pelton turbine will be expressed by the ratio of jet velocity to peripheral velocity and for Kaplan turbine by the ratio of flow velocity to peripheral velocity.

The kinematic similarity of different turbines can be expressed in terms of the known parameters by considering the following steps:

For Pelton turbine

$$Q = \frac{\pi}{4} d^2 v \tag{5.88}$$

Jet ratio, $M = \frac{D}{d}$, $d^2 = \frac{D^2}{M^2}$

$$Q = \frac{\pi}{4} \cdot \frac{D^2}{M^2} v$$

$$Q \propto D^2 v$$

$$v \propto \frac{Q}{D^2}$$

$$U = \frac{\pi DN}{60} \tag{5.89}$$

$$U \propto ND$$

$$\frac{Q_m}{N_m D_m^3} = \frac{Q_p}{N_p D_p^3} = \text{constant} \tag{5.90}$$

where, D is Diameter

d is jet diameter

Q is Discharge and

N is Revolution per minute (rpm)

For Francis turbine

$$Q = \pi DBv \tag{5.91}$$

$$\frac{B}{D} = k, B = kD$$

$$Q = \pi kD^2 v \tag{5.92}$$

$$Q \propto D^2 v \tag{5.93}$$

$$\frac{Q_m}{N_m D_m^3} = \frac{Q_p}{N_p D_p^3} = \text{constant} \tag{5.94}$$

For Kaplan turbine

$$Q = \frac{\pi}{4}(D^2 - d^2)v \qquad (k = \frac{d}{D}) \tag{5.95}$$

$$Q = \frac{\pi}{4} D^2(1 - k^2)v$$

$$Q \propto D^2 v$$

$$\frac{Q_m}{N_m D_m^3} = \frac{Q_p}{N_p D_p^3} = \text{constant} \tag{5.96}$$

From the expressions derived above it is seen that the condition for kinematic similarity for all type of turbine is same.

5.8.3 Dynamic Similarity

As per the dynamic similarity the ratio of forces at two different locations in one machine is same to that of another machine. The dynamic similarity in terms of known parameters can be derived as follows;

Dynamic force, F

$$F \propto \rho u^2 A$$

$$F \propto \rho D^2 N^2 L^2 \tag{5.97}$$

Static force, f

$$f \propto pA$$

$$f \propto QHL^2 \tag{5.98}$$

By taking Eq. (5.98) divided by Eq. (5.97), we get

$$\frac{H}{D^2N^2} = \text{constant} \tag{5.99}$$

$$\frac{H_m}{D_m^2N_m^2} = \frac{H_p}{D_p^2N_p^2} = \text{constant} \tag{5.100}$$

5.8.4 Specific Speed

The specific speed can be defined as the speed of turbine to develop unit power under unit head of water. It can be derived by considering the similarity law.

From kinematic similarity law,

$$\frac{Q}{D^3N} = \text{constant} \tag{5.101}$$

From dynamic similarity law,

$$\frac{H}{D^2N^2} = \text{constant} \tag{5.102}$$

Squaring Eq. 5.101, we get

$$\frac{Q^2}{D^6N^2} = \text{constant} \tag{5.103}$$

$(Eq. 5.102)^3$, we get

$$\frac{H^3}{D^6N^6} = \text{constant} \tag{5.104}$$

Dividing Eq. 5.101 by Eq. 5.104, we get

$$\frac{Q^2}{D^6N^2} \times \frac{D^6N^6}{H^3} = \text{constant} \tag{5.105}$$

$$\frac{Q^2N^4}{H^3} = \text{constant}$$

$$\frac{N\sqrt{Q}}{H^{3/4}} = \text{constant} \tag{5.106}$$

$$P \propto Q \times H$$

$$Q \propto \frac{P}{H} \tag{5.107}$$

Substituting the value of Q in Eq. 5.106, we get

$$\text{Specific speed, } N_s = = \frac{N\sqrt{P}}{H^{5/4}} \tag{5.108}$$

Example 5.12 A propeller turbine develops 950 kW under the head of 4.5 meters. A geometrically similar model of turbine, 1 meter in diameter develops 22 kW when running at 100 r.p.m. under the head of 2 meters. Determine the speed, diameter and specific speed of actual turbine.

Solution

Given data:

Model turbine	Actual turbine
$P_m = 22$ kW	$P_p = 950$ kW
$H_m = 2$ m	$H_p = 4.5$ m
$N_m = 100$ r.p.m	$N_p = ?$
$D_m = 1$ m	$D_p = ?$
	$(N_s)_p = ?$

Speed,

$$(N_s)_p = (N_s)_m$$

$$\frac{N_p\sqrt{P_p}}{H_p{}^{5/4}} = \frac{N_m\sqrt{P_m}}{H_m{}^{5/4}}$$

$$\frac{N_p\sqrt{950}}{4.5^{5/4}} = \frac{100\sqrt{22}}{2^{5/4}}$$

$$N_p = 41.93 \text{ r.p.m.}$$ **Ans.**

Diameter,

Diameter can be calculated based on dynamic similarity

$$\frac{H_m}{D_m^2 N_m^2} = \frac{H_p}{D_p^2 N_p^2}$$

$$\frac{2}{1^2 \times 100^2} = \frac{4.5}{D_p^2 \times 41.93^2}$$

$$D_p = 3.57 \text{ m.}$$ **Ans.**

Specific speed

$$(N_s)_p = \frac{N_p\sqrt{P_p}}{H_p{}^{5/4}}$$

$$= \frac{41.93 \times \sqrt{950}}{4.5^{5/4}}$$

$$= 197.183.$$ **Ans.**

EXAMPLE 5.12

5.9 UNIT QUANTITIES

Generally, there are five variable parameters; head (H), discharge (Q), speed (N), power (P), and efficiency (η) taken into account in order to study the performance of water turbines. Representation of discharge, speed and power parameters as unit quantity is considered good to discuss the performance of water turbines. Therefore, it becomes necessary to convert these parameters into unit quantities by following the steps discussed below:

Unit discharge (Q_u):

As the discharge is proportional to square of runner (D^2) and flow velocity (v_f), it is therefore the unit discharge (Q_u) can be expressed as;

$$\frac{Q_1}{Q_2} = \sqrt{\frac{H_1}{H_2}} \tag{5.109}$$

If H_1, is unity, then for unit discharge finally is expressed as;

$$Q_u = \frac{Q}{\sqrt{H}} \tag{5.110}$$

Unit speed (N_u):

Based on the concept discussed above, the relationship between head (H) and speed (N) can be expressed as;

$$\frac{H_1}{N_1^2} = \frac{H_2}{N_2^2} \tag{5.111}$$

If H_1 is unity, then for unit speed finally is expressed as;

$$N_u = \frac{N}{\sqrt{H}} \tag{5.112}$$

Unit power (P):

As the power is proportional to discharge (Q) and head (H) as expressed below;

$$P \propto H^{3/2} \tag{5.113}$$

$$\frac{P_1}{H_1^{3/2}} = \frac{P_2}{H_2^{3/2}} \tag{5.114}$$

If H_1 is unity, then for unit power finally is expressed as;

$$P_u = \frac{P}{H^{3/2}} \tag{5.115}$$

Example 5.13 A Francis turbine working under a head of 6 m at the speed of 200 rpm developed 74 kW when the rate of flow of water is 1.6 m^3/s. The runner diameter is 1.2 m. If the head on this turbine is increased to 17 m, determine its new speed, discharge and power.

Solution

$$N_1 = \frac{N\sqrt{H_1}}{\sqrt{H}}$$

$$= \frac{200 \times \sqrt{17}}{6} = 137.43 \text{ r.p.m. } \textbf{Ans.}$$

$$Q_1 = \frac{Q_1\sqrt{H_1}}{\sqrt{H}}$$

$$= \frac{1.6 \times \sqrt{17}}{\sqrt{6}} = 2.693\ m^3/s \textbf{ Ans.}$$

$$P_1 = \frac{P\,H_1^{3/2}}{H^{3/2}}$$

$$= \frac{74 \times 17^{3/2}}{6^{3/2}}$$

$$= 352.92 \text{ kW. } \textbf{Ans}$$

EXAMPLE 5.13

5.10 PERFORMANCE CURVES FOR WATER TURBINES

In order to know the characteristics of various water turbines their behaviour (performance) under variable conditions should be known. The curves which represent the water turbine performance are known as performance curves. The parameters considered to draw these curves are (i) head (H), (ii) discharge (Q), (iii), speed (N), (iv) power (P), (v) efficiency (η) and (vi) position of flow control mechanism (gate opening). The first three parameters (i to iii) are considered independent parameters. Further, as discussed under section 5.9 that it is better to represent the performance parameters in terms of unit quantities. Accordingly, based on the above-mentioned parameters the complete characteristics of water turbines are represented with the following curves;

5.10.1 Main Characteristic Curves

To draw these curves head (H) is kept constant and the variation of one parameter with respect to another parameter is considered. For example, the variation of discharge (Q), power (P) and efficiency (η) w.r.t speed (N) is shown for different water turbines in Fig. 5.28.

It can be seen in the figure that discharge (Q) drawn by the Pelton turbine remains constant w.r.t speed (N). Whereas for the Francis turbine discharge (Q) is decreased while in case of Kaplan turbine it increases for given value gate opening. The trend of power (P) and efficiency (η) w.r.t speed (N) for given gate opening (G.O.) remains similar for all the water turbines.

Basically, these curves are useful to the designers during design stage of water turbines.

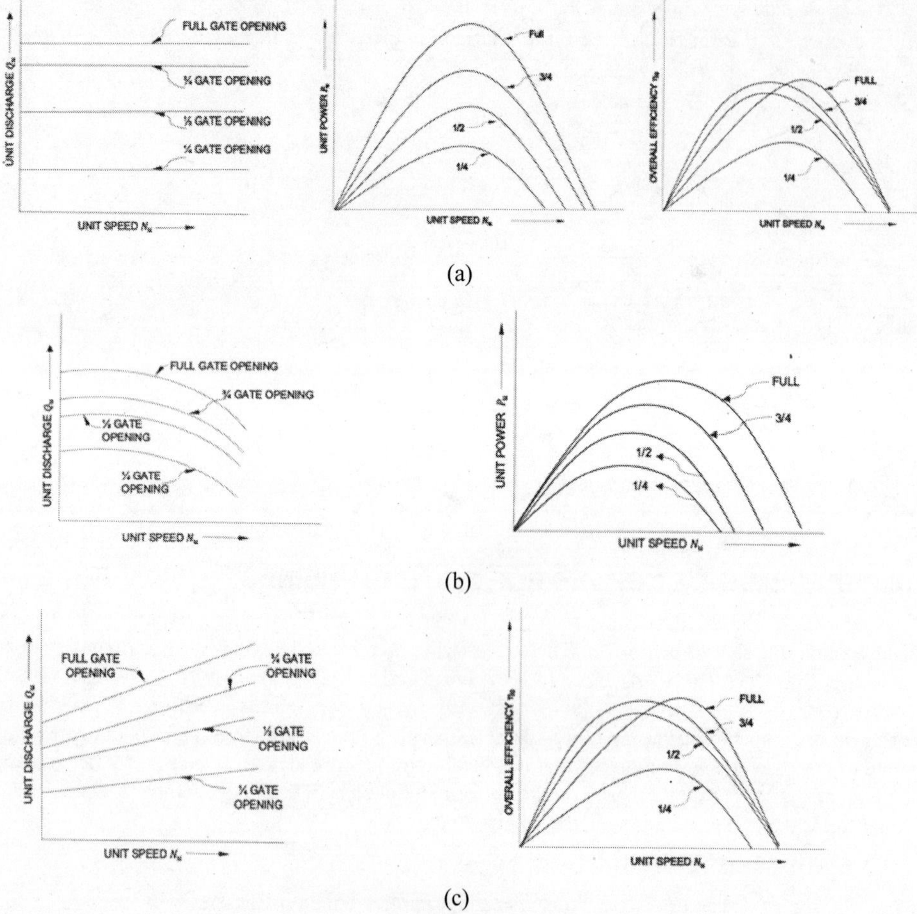

Fig. 5.28: Main characteristics curves for (a) Pelton turbine (b) Francis turbine (c) Kaplan turbine

5.10.2 Operating Characteristic Curves

Operating characteristic curves are useful for the hydro power station developers or to the operators. In a working hydro power station, the developers/operators need to observe the turbine efficiency w.r.t discharge variation under different operating conditions of water turbines. Actually, the water turbine operating conditions is to run the turbine at constant speed to derive the generator in order to maintain the frequency under different load conditions in a hydro power station. It is therefore, the discharge is varied correspond to different loads over turbine to maintain the speed constant. With this concept, the operating characteristics are drawn between efficiency (η) versus discharge (Q) for constant speed corresponding to given head (H) is shown in Fig. 5.29.

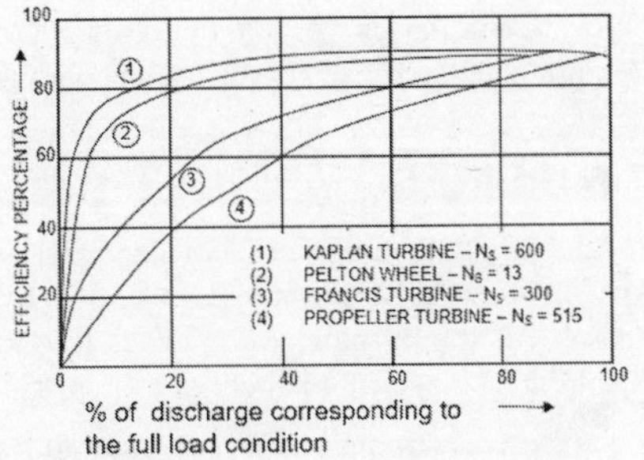

Fig. 5.29: Operating characteristic curves

5.10.3 Constant Efficiency Curves

Constant efficiency curves are also known as hills curves, iso-efficiency or Muschel curves as shown in Fig. 5.30. These curves normally used for commercial purposes as sometimes the guarantee efficiency given by the supplier becomes to validate under actual site conditions by the hydro power station developers. The head and discharge conditions at which the efficiency guaranteed by the supplier may be different under actual site conditions after installation of the turbine. Then the efficiency obtained under actual site conditions can be verify from these curves which has to be agreed by both parties i.e., developer and supplier.

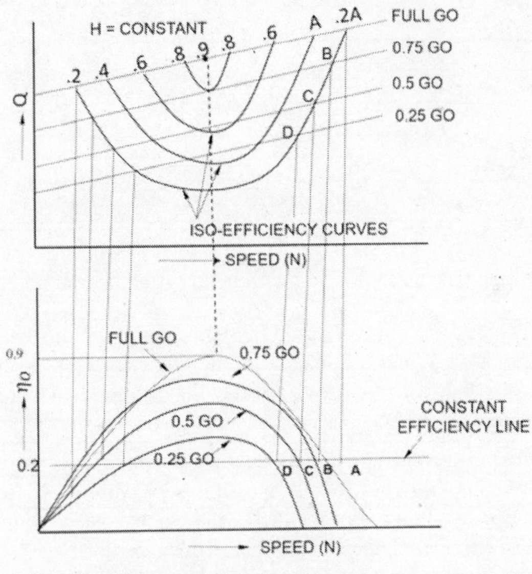

Fig. 5.30: Iso-efficiency curves

Basically, these curves are drawn from the data generated during testing of water turbines for preparation of main characteristic curves. For example, the data obtained to draw discharge variation with speed and efficiency versus speed are used to draw the operating characteristic curves for Kaplan turbine as shown in Fig. 5.28 (c). Similar curves for Pelton and Francis turbine can be drawn by considering their respective main characteristic curves as shown in Figs. 5.28 (a) and 5.28 (b) respectively.

5.11 GOVERNING OF TURBINES

Governing of water turbines means to run the turbines at constant speed under different load conditions in the hydro power stations. Conventionally, discharge is varied by controlling the flow control mechanism (control valve in case of Pelton turbine, guide vanes for Francis turbine, and wicket gates and runner blades in case of Kaplan turbine) with the help of system known as governor.

Fig. 5.31. shows schematic of a typical governing system for water turbines in a hydro power station. It consists of following main components;

(i) Speed of turbine/generator frequency sensing device

(ii) Servo motor

(iii) Pressurized oil system

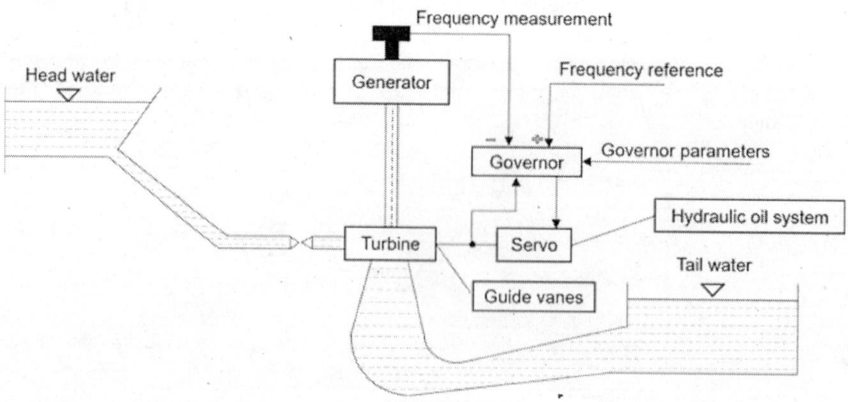

Fig. 5.31: Schematic showing the governing system for water turbines

The working of a simple oil pressure governor used in hydro power station is described with respect to a schematic shown in Fig. 5.32. When the load over the turbine increases the speed will decrease corresponding to the given discharge thereby governor flyballs moves down words (1) results in to move up the other end of the floating lever (2).

This downward movement of the lever end will push down pilot which will connect the pressurize oil pipe to the upper end of the piston of the servo motor (3). The piston connected to the flow control mechanism which will be opened accordingly with movement of the piston rod as indicated the gates open in the Fig. 5.32. Similar procedure in reverse mode will be performed if the load gets reduced over machine and speed (*N*) will increase. As shown in the figure, an oil dashpot is used to adjust the position lever with respect to the gate opening (4) for at rated load and rated speed (*N*) of the turbine.

SCAN ME

For better understanding governing of turbine

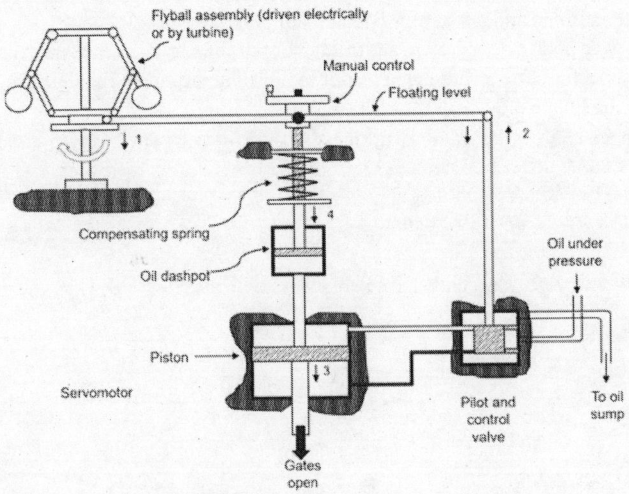

Fig. 5.32. Oil pressure governing system

5.12 SELECTION OF TURBINES

As a thumb rule, for high heads (say $H > 300$ m), Pelton turbines are used, and for the low head ($H < 25$ m), one or the other variant of Kaplan turbine is used. In the gap between these two turbines and over lapping with Pelton on one side and Kaplan on the other, Francis turbines are extensively used.

Scientifically, the selection of hydro turbines is made based on specific speed (N_S) as expressed by Eq. 5.116.

$$\text{Specific speed } (Ns) = \frac{N\sqrt{P}}{H^{5/4}} \tag{5.116}$$

Specific speed is measured in metric unit, where N is speed in rpm, P is turbine capacity in kW and H is head in m. The range of specific speed for different turbines are given in Table 5.6.

Table 5.6: Range of specific speed for different water turbines

Types of turbines	Specific speed (N_S)
Pelton	12 – 35 for single jet 35- 85 for multi jet.
Francis	80 – 400
Propeller and Kaplan	340 – 1000

It can be seen from the table that the maximum value of specific speed is 1000. It is therefore, it's not possible to build water turbine having specific speed more than 1000. It is recommended while calculating the specific speed for given data its value should be less than 1000. There is some flexibility regarding the selection of specific speed by assuming speed of turbine and number of units for a given capacity of hydro power station. In case of two different machines are suitable for a given site due to overlapping of specific speed range, the turbine with lower specific energy cost has to be chosen. In case if low-speed turbine is selected to keep the specific speed in the desired range, speed increaser will be introduced between turbine and generator to match the available speed of generator. Further, for large capacity of hydro power plant under low head, if it is not possible to keep the specific speed value less than 1000 either by selecting low speed generator or by introducing speed increaser then the last option adopted is to divide the total capacity in number of units.

EXAMPLE 5.14

Example 5.14 Select the suitable turbines for the following sites having various capacities under different heads

(i) $P = 100$ kW at $H = 38$ m

(ii) $P = 150$ kW at $H = 150$ m and

(iii) $P = 400$ kW at $H = 3$ m

The turbine is required to be coupled to a generator having 1500 rpm if possible.

Solution

$P = 100$ kW at $H = 38$ m

Specific speed,

$$N_s = \frac{1500\sqrt{100}}{38^{5/4}}$$

$N_s = 158.9.$

Therefore, Francis turbine is suitable. **Ans.**

$P = 150$ kW at $H = 150$ m

Specific speed

$$N_s = \frac{1500\sqrt{150}}{150^{5/4}}$$

$N_s = 35.$

Hence, Single jet Pelton turbine is suitable. **Ans.**

$P = 400$ kW at $H = 3$ m

Specific speed,

$$N_s = \frac{1500\sqrt{400}}{3^{5/4}}$$

$N_s = 7598$, which is much higher than the specific speed limit of 1000.

In such case, there is only option for selecting specifically designed low rpm generator and increase number of units accordingly to have the feasible solution for this case generator of 300 rpm and 4 number of units is suggested. Then, the specific speed for these selected parameters is determined as

$$N_s = \frac{300\sqrt{\frac{400}{4}}}{3^{5/4}}$$

$N_s = 760$

Hence, Kaplan turbine is suitable. **Ans.**

Example 5.15 A power plant is to be built on a river having flow rate of 67 cumec. and utilizing 12.8 m of head. The speed is fixed by the electrical company is 167 rpm. The efficiency of the turbine is 90%. It is further seen that the specific speed of 342 would be best for the hydro plant. Selection of the turbine is to be made. Find out (i) Type of turbine (ii) Number of turbines required.

Solution

Given data

$Q = 67\ m^3/sec$, $H = 12.8\ m$, $N = 167$ rpm, $\eta = 90\ \%$

Best N_s for hydro power plant = 342

Efficiency, $\eta = \frac{P}{\rho g Q H}$

Power, $P = \eta \times \rho g Q H$

$$P = 0.9 \times 10^3 \times 9.81 \times 67 \times 12.8$$

$$P = 7571.75\ \text{kW}$$

Specific speed, $N_s = \frac{N\sqrt{P}}{H^{5/4}}$

$$N_s = \frac{167\sqrt{7571.75}}{(12.8)^{5/4}}$$

$$N_s = 600\ (does\ not\ match\ with\ 342)$$

Now, select number of turbines = 2

$$N_s = \frac{167\sqrt{(7571.75)/2}}{(12.8)^{5/4}} = 424.2\ \neq 342$$

Now, select number of turbines = 3

$$N_s = \frac{167\sqrt{(7571.75)/3}}{(12.8)^{5/4}} = 346.5\ \neq 342$$

(i) Type of turbine = Francis turbine ($N_s = 80\ to\ 400$). **Ans.**

(ii) Number of turbines = 3 **Ans.**

EXAMPLE 5.15

UNIT SUMMARY

- ***Working principle of water turbines***

  According to Newton's Second Law

  Dynamic force = rate of change in momentum

$$F = m.\frac{dv}{dt}$$

Force exerted:

Stationary flat plate, $F = \rho a v^2$

Inclined flat plate, $F_x = F \sin\theta$

$$= \rho a v^2 \sin^2\theta$$

Moving flat plate, $F = \rho a(v-u)^2$

Work done by the jet

On plate, $F = \rho . a(v-u)^2 . u$

Series of flat plate, $W = F . u = \rho\, a\, v\,(v-u)u$

On curved stationary plate in x-direction:

$$F_x = \rho . a\, v_1\, (v_1 \cos\alpha_1 - v_2 \cos\alpha_2)$$

On moving curved plate:

$$W = \rho a(v_1 - u)(v_1 \cos\alpha_1 - v_2 \cos\alpha_2)$$

- ***Head and Efficiencies***

Net Head, $H_{\text{net}} = H_{\text{gross}} - h_l$

Hydraulic efficiency, $\eta_h = \dfrac{\text{power produced by the runner}}{\rho g Q H}$

- ***Pelton turbine***

Suitable for high head and low discharge

Diameter of runner, $D = \dfrac{38.9\sqrt{H}}{N}$

Jet velocity, $v_{jet} = K_v\sqrt{2gH}$

Runner velocity, $u = \dfrac{\pi D N}{60}$

Breadth of bucket, $B = (2.8 - 3.2)d$

Height of bucket, $H = (2.2 - 2.8)d$

Depth of bucket, $T = (0.6 - 0.9)d$

Number of buckets, $Z = 0.5\,m + 15$

Kinetic energy of the jet = $\frac{1}{2}\rho a v_1^3$

Hydraulic efficiency (η_h):

$$\eta_H = \frac{2(v_1 - u)[1 + \cos\varphi]u}{v_1^2}$$

For efficiency to be maximum

$$\text{u} = v_1/2$$

Discharge of the jet through n_j nozzles:

$$q = \frac{\pi}{4} d_j^2 \times K_v(\sqrt{2gH}) \qquad (q = \frac{Q}{n_j})$$

Diameter of jet, $d = \dfrac{0.54\sqrt{\frac{Q}{n_j}}}{H^{\frac{1}{4}}}$

Design of casing:

Width = $(12 - 18)d$

Width at upper cover = $(4 - 5)d$

Height of casing = $(2-3)D = (24-36)d$

- ***Francis turbine***

Suitable for medium head and medium discharge

Sprouting velocity, $v = \sqrt{2gH}$

The discharge at different section of the casing

$$Q_\theta = Q\,\frac{\theta}{2\pi}$$

Flow ratio, $\psi = \frac{v_f}{\sqrt{2gH}}$

Speed ratio, $K_u = \frac{u}{\sqrt{2gH}}$

Runner outlet diameter, $D_o = \frac{60}{\pi N} K_u \sqrt{2gH}$

Runner inlet diameter, $D_i = D_o/2$

Number of runner vanes, $Z_1 = Z + 1$

Runner width taking blade thickness into consideration

$$Q = (\pi\,\mathrm{D} - Z_1\mathrm{t})B\,v_f$$

Efficiency of draft tube, $\eta_d = \frac{\left[\frac{v_2^2 - v_3^2}{2g}\right] - h_f}{\frac{v_2^2}{2g}}$

Design of draft tube:

Diameter at the inlet of the draft tube (D_{di}) = outlet diameter of the runner (D_o)

Semi cone angle α = 4 to 8°

Length of the draft tube $L_D = D_2(5 - \frac{N_s}{200})$

Depth of the draft tube $H = D_2(3.4 - \frac{N_s}{400})$

- ***Kaplan turbine***

Suitable for low head and high discharge

The efficiency of a draft tube is, $\eta_d = \frac{\left[\frac{v_2^2 - v_3^2}{2g}\right] - h_f}{\frac{v_2^2}{2g}}$

Speed ratio, $K_u = \frac{u}{\sqrt{2gH}}$

Flow ratio, $\psi = \frac{v_f}{\sqrt{2gH}}$

Diameter of the runner, $D_r^2 = \frac{4Q}{\pi v_f[1-k^2]}$ where, $k = \frac{D_h}{D_r}$

Efficiency of Kaplan turbine

$$\eta = \frac{(v_{u1} + v_{u2})u}{gH}$$

Design of wicket gates:

Diameter, $D_o = D_r \times K_{u1}$

Width, $B = 0.4\, D_r$

Design of casing:

$v = 0.05 - 0.01\sqrt{2gH}$

$Q = 2.5D \times h \times v$

$$h = \frac{Q}{2.5D \times 0.075\sqrt{2gH}}$$

- ***Similarity law***

Geometrical similarity, $\left(\frac{B}{D_1}\right)_m = \left(\frac{B}{D_1}\right)_p$ = constant

Kinematic similarity, $\left(\frac{v_f}{u}\right)_m = \left(\frac{v_f}{u}\right)_p$ = constant

Dynamic similarity, $\frac{H_m}{D_m^2 N_m^2} = \frac{H_p}{D_p^2 N_p^2}$ = constant

Specific Speed, $\frac{N\sqrt{P}}{H^{5/4}}$

- ***Unit quantities***

Unit discharge, $Q_u = \frac{Q}{\sqrt{H}}$

Unit speed, $N_u = \frac{N}{\sqrt{H}}$

Unit power, $P_u = \frac{P}{H^{3/2}}$

- ***Performance curves for water turbines***

Main characteristic curves:

Head (H) is kept constant and variation of one parameter w.r.t another parameter is considered.

Useful to the designers during design stage of water turbines.

Operating characteristic curves:

Discharge is varied correspond to different loads over turbine to maintain the speed constant

Useful for the hydro power station developers or operators to observe the turbine efficiency w.r.t discharge variation under different operating conditions.

Constant efficiency curves:

These curves are drawn from the data generated during testing of water turbines.

Efficiency obtained under actual site conditions can be verify from these curves.

- ***Selection of turbines***

Types of turbines	Specific speed (N_S)
Pelton	12 – 35 for single jet 35- 85 for multi jet.
Francis	80 – 400
Propeller and Kaplan	340 – 1000

EXERCISES

Multiple Choice Questions

5.1 Water turbine is the prime mover used to convert

(a) Electrical power to Mechanical power (b) Mechanical power to Electrical power

(c) Hydro potential to Mechanical power (d) (b) and (c)

5.2 Working principle of water turbines is based on

(a) Newton's 3rd law of motion (b) Impulse-momentum equation

(c) Bernoulli's equation (d) (b) and (c)

5.3 The maximum possible efficiency for a simple turbine where jet is striking number of flat plates mounted on the periphery of a wheel

(a) 30 % (b) 90 % (c) 50 % (d) 75 %

5.4 The ratio of peripheral velocity to jet velocity, to get maximum efficiency of Impulse turbine is

(a) 0.25 (b) 1 (c) 0.5 (d) 0.75

5.5 The function of casing in Pelton turbine is to

(a) Convert potential energy into kinetic energy (b) To avoid splashing of water

(c) To provide safe guard to the operator (d) (b) and (c)

5.6 The range of speed ratio for Francis turbine is

(a) 0.46 to 0.60 (b) 0.60 to 0.90 (c) 0.90 to 2.0 (d) 2.0 to 5.0

5.7 Flow ratio for a Reaction turbine is defined as

(a) Peripheral velocity to jet velocity (b) Jet velocity to peripheral velocity

(c) Flow velocity to spouting velocity (d) Spouting velocity to flow velocity

5.8 The functions of the draft tube are

(a) To facilitate the reaction turbine installation at higher level from tail race water level

(b) To facilitate the Impulse turbine setting at higher level from tail race water level

(c) To act as an energy recuperater for the energy going as waste at the exit of the runner

(d) (a) and (c)

5.9 In Kaplan turbine

(a) Blades are fixed

(b) Wicket gates are fixed

(c) Blades are fixed and wicket gates are adjustable

(d) Both blades and wicket gates are adjustable

5.10 In bulb turbine

(a) Generator is mounted on the outside of water passage

(b) Turbine and generator are encased inside the water passage

(c) (a) and (b)

(d) None of the above

5.11 The part load efficiency is better for

(a) Pelton turbine (b) Francis turbine (c) Propeller turbine (d) Kaplan turbine

5.12 For high head and low discharge hydro plant mostly, turbines are suitable

(a) Kaplan turbine (b) Francis turbine (c) Pelton turbine (d) (a) and (b)

5.13 If prototype of water turbine is designed based on similar model and efficiency of the model is 90% then the efficiency of prototype will be

(a) 85 % (b) 75 % (c) 90 % (d) 60 %

5.14 The specific speed (N_s) of water turbine is specified by

(a) $\frac{N\sqrt{P}}{H^{5/4}}$ (b) $\frac{N\sqrt{P}}{H^{3/2}}$ (c) $\frac{P\sqrt{N}}{H^{5/4}}$ (d) None of the above

5.15 Two water turbines can be said similar if they are

(a) Geometrical similar (b) Dynamic similar (c) Kinematic similar (d) All of the above

5.16 The range of specific speed of Francis turbine is

(a) 15 – 30 (b) 80 – 400 (c) 360 – 1000 (d) 1000 – 1500

5.17 Unit power is expressed by

(a) $P_u = \frac{P}{H}$ (b) $P_u = \frac{P}{H^{3/2}}$ (c) $P_u = \frac{P}{H^{1/2}}$ (d) $P_u = \frac{P}{H^{4/5}}$

5.18 The operating characteristic curves are drawn between

(a) Power versus Speed

(b) Head versus Efficiency

(c) Discharge versus Efficiency

(d) All of the above

5.19 In conventional hydro power stations governing is done by maintain the constant speed of turbine by controlling

(a) Load (b) Head (c) Discharge (d) (b) and (c)

5.20 The suitable turbine for a hydro power plant having the capacity of 100 kW, head as 1 m and speed as 100 r.p.m. is

(a) Francis turbine (b) Kaplan turbine (c) Propeller turbine (d) (b) and (c)

Answers of Multiple-Choice Questions

1.1 (c), 1.2 (b), 1.3 (c), 1.4 (c), 1.5 (d), 1.6 (b), 1.7 (c), 1.8 (d), 1.9 (d), 1.10 (b), 1.11 (d), 1.12 (c), 1.13 (c), 1.14 (a), 1.15 (d), 1.16 (b), 1.17 (b), 1.18 (c), 1.19 (c), 1.20 (b)

Short and Long Answer Type Questions

5.1 Classify water turbines according to action and direction of water over runner.

5.2 Derive the fundamental equation of hydro turbines.

5.3 Discuss the forces which play the role in working principle of a water turbine.

5.4 Name the main components of typical single jet Pelton turbine. Sketch the runner of all Pelton turbine and discuss the shape of Pelton turbine bucket with the help of a sketch in three views.

5.5 Define the term 'Governing of turbine'. Describe with neat sketch the working of an oil pressure governor.

5.6 What are the characteristics curves of water turbines? Draw the 'best efficiency curves' with the help of main characteristic curves for typical case of water turbines? Discuss the operating characteristics curves for Kaplan turbine with respect to Pelton turbine and Francis turbine.

5.7 Why a draft tube is advantageous in reaction turbines only and not in Pelton turbines? Derive the expression for draft tube efficiency.

5.8 Show the various components of a Kaplan turbine with suitable sketch.

5.9 Discuss the law of similarity for water turbines. Define specific speed for a water turbine and show that the specific speed for a model and actual machine must be same for similar conditions.

5.10 Define the terms "head and discharge" and "rated power and efficiency" as used in water turbines.

5.11 Discuss the working proportions for a Kaplan turbine? How these are different from the working proportions of Francis turbine?

5.12 What is the function of Pelton turbine casing and give the reason for not having its function in hydraulic performance?

5.13 Draw a sketch for Francis turbine showing all its main components. Discuss the functions of a draft tube and write the expression for efficiency of draft tube.

Numerical Problems

5.14 Using the velocity triangles, determine the power developed by a curved vane having deflection angle of 165° and moving with velocity of 15 m/s, if a water jet with a velocity of 35 m/s impinges to the curved vane. Considered the discharge of 170 lps.

5.15 Using the following data determine the (i) jet numbers (ii) jet diameter (iii) runner diameter (iv) dimensions of buckets and their numbers (v) discharge for a Pelton turbine.

A Pelton wheel has to be designed for the following data;

Capacity of the plant	: 5 MW
Head	: 250 m
Turbine speed	: 500 rpm
Jet ratio	: 10
Overall efficiency	: 0.85

5.16 Design a Pelton turbine for a site having a head of 60 m. The turbine is to develop 100 kW power at 200 rpm. Considering overall efficiency as 0.85 and assuming suitable values for required terms.

5.17 A power house is equipped with Pelton turbine which delivers a maximum power of 14500 kW under a head of 855 m and running at 600 rpm. Find the least diameter of the jet and the mean diameter of the wheel. Determine the value of the jet ratio and suggest whether is it within the safe limit? Also, find out the number of buckets with bucket dimensions for the Pelton turbine. Consider the overall efficiency as 85%.

5.18 Draw the velocity triangles for the flow over a Pelton bucket and define all the terms used. Water from a nozzle in the form a jet impinges on a series of hemispherical cups and is deflected through 180o. If the velocity of water jet is 30 m/s and that of the cups is 12 m/s, find the work done per second by the water if nozzle diameter is 7.5 cm.

5.19 Design the Pelton turbine with the following data;

Head of water = 300 m. (ii) Flow rate = 240 lps.

Speed of turbine = 750 rpm.

Assume the values of constants and coefficients required in calculations.

5.20 A site having power plant capacity of 368.0 kW under a head of 70 m. Determine the following parameters related to a Francis turbine (i) runner diameters (ii) runner width and (iii) discharge. Use in the calculations, turbine speed as 750 rpm, width to diameter ratio as 0.1, flow ratio as 0.15, and efficiency as 95%.

5.21 Determine the speed and power for a prototype of Francis turbine, if the model to prototype ratio 1:5. The model capacity is 3 kW at the speed of 300 rpm for a given head of 1.90 m. What will be the capacity and speed of the prototype? Assume in the calculations the head for the prototype 6 m, efficiency for the model and prototype is same as 76%.

5.22 The following data is given for a Francis turbine. Net head, H =60 m; Speed, N = 700 rpm; Output power, P = 295 kW; overall efficiency = 84%; Flow ratio = 0.20; Breadth ratio = 0.1; Outer diameter of the runner = 2 x inner diameter of the runner. The thickness of the runner occupies 5% of the circumferential area of the runner. Determine:

i) Diameters of runner at inlet and outlet, and

ii) Width of runner at inlet.

5.23 Design a Francis turbine runner and find out the speed of runner with the following data: Net head, H = 65m, Output power, P= 325 kW, Overall efficiency, η_o = 86%, Flow ratio, ψ = 0.14, Breadth to runner diameter ratio, m =0.11, Inner diameter of runner is 0.5 times of outer diameter and speed ratio k_u =0.65.

5.24 A Kaplan turbine develops 5 MW at a head of 6 m. Its speed ratio is 2 and flow ratio is 0.6 and the diameter of the hub is 0.35 times the outer diameter of the runner. The efficiency of the turbine is 0.90. Calculate the diameter of the runner, speed of the runner and also its specific speed.

5.25 The design parameters for a model are as head of 6 m and efficiency as 90%. Calculate the design parameter of a Kaplan turbine for a capacity of 10.25 MW under a head of 9.5 m and rpm of 120. Considering the model to prototype scale as 1:10.

5.26 Design a Kaplan turbine runner to develop 8.5 MW under a head of 5.5m. Assume, the speed ratio as 2.1, flow ratio as 0.65, overall efficiency as 90% and hub to runner diameter as 1:3. Also, find out the speed and specific speed of the turbine.

5.27 A Propeller turbine develops 800 kW under the head of 5.5 m. A model of the turbine, 1 m in diameter develop 15 kW when running at 120 rpm under the head of 2 m. Determine the speed, diameter and specific speed of actual turbine.

KNOW MORE

History of Water Mill

The idea of using energy in water and converting it into mechanical energy was known to the mankind since long. In ancient times, the mechanical power was developed by passing flowing water through wheels, such wheels known as water wheels were used in China, India, Egypt and later in Europe. Traditional water mills for grinding are being widely used in the Himalayan regions and about 2.5 lacs traditional wooden water mills are still in use in Himalayan and Sub-Himalayan regions. These are vertical shaft wooden mills generally used for grinding grains only and are called as "Gharat". These gharats operate under a head of 2 to 5 m to produce an average of 0.5 kW mechanical outputs. As the requirement of energy increased for the development of society the best form of energy is considered as electricity. Accordingly, this water mills have been converted into improved devices known as Water turbine to derive the generator for electric generation.

Applications

Among all the renewable energy sources, hydropower is considered as one of the best renewable energy source which is available in water streams at different locations like hills, canals in planes. To convert this hydro potential into electricity a hydro power plant has to be established. A typical hydro power plant consists of various components, out of them water turbine is considered as heart of the pant. Depending upon the location of potential sites, different types of turbine are used covering different range of head and discharges.

The water turbine technology is well matured technology with the highest conversion efficiency among all prime movers. Globally, hydro power contributes the maximum share in electricity generation compared to all the renewable sources. As far as enhancement in efficiency of water turbine concerned there is hardly any scope however, there are some issues related to operation and maintenance of water turbines. Researchers are working to mitigate such challenges.

REFERENCES AND SUGGESTED READINGS

List of some of the books is given below which may be used for further learning of the subject:

1. E. Mosonyi, Water Power Development, Vol. I and II, Nem Chand and Brothers, 2009.
2. P.S. Nigam, Handbook of Hydroelectric Engineering, Nem Chand and Brothers, 2001.
3. J. Lal, Hydraulic Machines, 3rd edition (reprint), Metropolitan Book Co. Private Limited, 2002.
4. National and International Standards
5. G. Brown, Hydro-electric Engineering Practise, Vol. II, CBS Publication, 1984.

REFERENCES FOR FURTHER LEARNING

1. R.J. Garde and A.G. Mirajgaoker, Engineering Fluid Mechanics, SCITECH Publication, 2010.
2. Streeter & Wylie, Fluid Mechanics, Tata McGraw Hill, 9th edition.
3. K. Subramanya, Fluid Mechanics and Hydraulic Machines, Mc Graw hill, 2018.
4. Frank White, Fluid Mechanics, Mc Graw Hill, 7th edition.
5. J. Guthrie Brown, Hydro-Electric Engineering Practice, CBS Publishers & Distributors, 2nd edition.

CO AND PO ATTAINMENT TABLE

Course outcomes (COs) for this course can be mapped with the programme outcomes (POs) after the completion of the course and a correlation can be made for the attainment of POs to analyze the gap. After proper analysis of the gap in the attainment of POs necessary measures can be taken to overcome the gaps.

Table for CO and PO attainment

Course Outcomes	Attainment of Programme Outcomes *(1- Weak Correlation; 2- Medium correlation; 3- Strong Correlation)*											
	PO-1	PO-2	PO-3	PO-4	PO-5	PO-6	PO-7	PO-8	PO-9	PO-10	PO-11	PO-12
CO-1												
CO-2												
CO-3												
CO-4												
CO-5												

The data filled in the above table can be used for gap analysis.

INDEX